ANATOMY AND ACTIVITIES OF PLANTS

A guide to the study of flowering plants

C. J. Clegg Ph.D., DIC, M.I.Biol.
Avery Hill Campus, University of Greenwich

Gene Cox
Managing Director, Microcolour Ltd

JOHN MURRAY 50 Albemarle Street London

Preface

All animals derive their food either directly or indirectly from plants, and consequently the study of plant life is central and fundamental to an understanding of the living world. For mankind one of the major world problems is finding the means to feed the growing population of the world, and much emphasis is placed on improvements in food production by growing better crop plants. For these reasons the need for sound factual knowledge of plant life is important for all mankind.

Anatomy and Activities of Plants is a pictorial account of the cells and tissues of flowering plants, and their functions. It shows where tissues occur in mature organs, it outlines the development of tissues and it is concerned with the ways in which the structure and arrangement of tissues facilitates the functioning of the whole plant.

It has been assumed that the reader will have studied basic biology, and will already be familiar with the cell concept and with the gross structure of the plant. Included in the section on books for further reading are some examples of introductory books for background study. Readers who have leisure interests concerning plants or students whose botanical knowledge is not advanced may like to use this book together with introductory ones. Inside the front cover is a Glossary of terms that are used in the text. It provides explanations and reminders, rather than formal definitions.

The number of young people in schools and colleges who study biology at an advanced level is increasing. For these students this book anticipates the sort of questions that arise in study of plant histology, and relates the detailed structure to gross anatomy and to major physiology topics so that the reader maintains an idea of the whole functional plant.

Many of the photographs that have been selected to illustrate structure are linked to drawings identifying the plane of the section or the position of the tissue in the plant. Some photographs, too, are accompanied by tissue map drawings that record the relative proportions of tissues, simply and accurately, without the drawing of individual cells. A few drawings show how details of individual cells can be recorded by representing one or two cells of each type, highly magnified. All these drawings may serve as models for the students' own work. The combined effect of photographs, drawings and text contributes to the building of skills in interpretation, and of recording of detail of the structure and functioning of plants.

C.J.C.
G.C.

Acknowledgements

The authors would like to thank the following who read the manuscript and made many constructive criticisms. We greatly value the contribution they have made to the accuracy and clarity of the text: Mr. Don Mackean, Dr Richard Johnson, Dr Cecil Prime and Mrs Pat Winter.

The authors are indebted to the following people who have provided additional photographic material: Heather Angel, 2.10(i), p.63; Dr Alan Freundlich, 1.3(iv); Philip Harris (Biological) Ltd, 3.6(A), 3.6(B), 3.6(C); Howard Jay, 1.20(iii), 1.21(i), p.64 (Bracken, Scots Pine); Dr Richard Johnson, 4.3(i), 4.10(ii); ARDEA (John Mason), p.64 (Club Moss); P. Wakely, 1.24(i). The authors also thank Mrs E. Taylor and Mr K. Poole for much help in the preparation of materials.

Dr Clegg is particularly grateful to Mr Don Mackean for his advice in the early stages of making the drawings.

Gene Cox would also like to thank T. Gerrard & Co, for providing some of the material to photograph; also Mrs Mills, Biology Department, Royal Free Hospital School of Medicine, and Dr M Rothschild both for the loan of specimens and for much help and advice.

© C.J. Clegg and Gene Cox 1978

First published 1978

Reprinted 1980, 1986, 1988, 1990,1992,1994,1996, 2000

Printed by Colorcraft Ltd.

0 7195 3319 8

Contents

Companion book

Lower Plants: Anatomy and Activities of Non-flowering Plants and their Allies,
C. J. Clegg. John Murray, 1984.

This book is a similar, concise, pictorial account of the complete range of lower plants, and their allies studied at A level and similar courses. Structural, physiological, biochemical, and ecological aspects are covered, along with essential features of taxonomy, reproduction and life-cycle.

Books for further reading

1. Background study for beginners:

GCSE Biology, second edition, D.G. Mackean (1995). John Murray. Section Two, pages 52–92.
Biology for Life, M.B.V. Roberts (1986). Nelson. Pages 42–3 & 220–53.

2. Texts on Biology at Advanced level:

Advanced Biology; Principles & Applications, C.J. Clegg with D.G. Mackean (1994). John Murray.
Biology – A Functional Approach, M.B.V. Roberts (1986). Nelson.

3. Advanced texts on the anatomy of plants:

Anatomy of Seed Plants, K. Esau (1977). John Wiley & Sons.
Plant Anatomy, A. Fahn (1993). Pergamon.

4. Advanced texts on the physiology of plants:

The action plant, P. Simons (1992). Blackwell.
Plant Physiology, F. Salisbury & C. Ross (1992). Wadsworth.

5. Texts on evolution and life cycles of plants:

Green Plants; Their Origin and Diversity, P.R. Bell (1992). CUP.
Diversity and Evolution of Land Plants, M. Ingrouille (1992). Chapman & Hall.

6. Economic importance of plants:

Plant Science, An Introduction to World Crops, J. Janick, R. Schery, F. Wood and V. Ruttan (1981). Freeman.

7. Pollination of flowers:

The Pollination of Flowers, M. Proctor and P. Yeo (1973). Collins.

8. Ecology:

Techniques and Fieldwork in Ecology, G. Williams (1987). Bell & Hyman.

Section 1: The stem - support and transport

Introduction

The **stem** supports the leaves and flowers in the sunlight, transports organic materials (e.g. sugar, amino acids), ions, and water between the roots and the aerial system, and it is the site of some starch storage.

The stem originates from the *plumule* of the seed (see page 62), and bears leaves attached at points called *nodes*. At the top of the stem is a *terminal bud* or terminal growing point, and in the axil of each leaf a *lateral bud* or lateral growing point (see page 38). New cells are produced at growing points. These are regions of active cell division and are called *meristems*. *Apical meristems* occur at the apex of stem (main stem and lateral stems) and roots. The activity of apical meristems causes growth in length of stem and root. *Lateral meristems* occur in older parts of plants, situated parallel to the sides of the plant, and their activity causes increase in thickness of stem and root.

The cells formed from the apical meristem differentiate to become the primary growth of the plant. Herbaceous (non-woody) plants are largely or only formed by primary growth. Herbaceous stems tend to be relatively short-lived, and some herbaceous plants survive the unfavourable season (e.g. winter) as seeds or as underground organs. When lateral meristems become active, secondary growth occurs and a woody structure results, protected by bark. The trees and shrubs formed by secondary growth have stems that survive for many years.

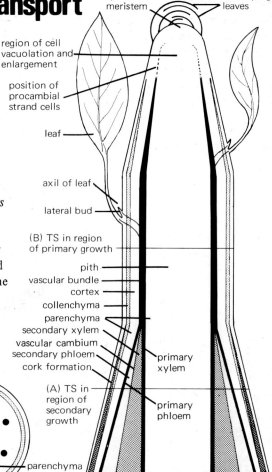

Figure 1.1(i) Diagrammatic representation of a growing stem shown in longitudinal section (LS) and transverse sections (TS)

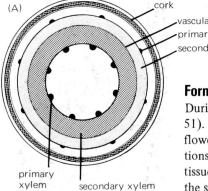

Formation of the stem

During periods of active growth of the plant the cells of the meristem divide by *mitosis* (see page 51). The cells cut off from the meristem differentiate into the mature cells of the plant. The flowering plant is a multicellular organism in which the cells are specialized for particular functions. A group of cells having the same function are known as a *tissue*. There are several kinds of tissue grouped together to form organs such as the stem and the leaf. In the primary growth of the stem the following tissues are formed.

Tissues formed in primary growth	Where they occur in the stem	What functions the tissue has	Turn to these pages for more information
epidermis	outermost layer, one cell thick	protects the stem contains all the cells under pressure within one continuous skin and thereby contributes to supporting the plant	page 10 and page 32 (epidermis of leaves)
parenchyma	fills the greater part of the cortex and pith	shows relatively little specialization, may be concerned with various physiological functions of the plant; its turgidity causes pressure on the epidermis which creates support for the stem	page 10
collenchyma	often immediately below the epidermis in young stems	supporting tissue in young growing stems and in leaves	page 10
fibres	in and around vascular bundles, in the cortex of mature stems	support and protection	page 9
xylem vessels	in the vascular bundle	transport of water	page 6
tracheids	in the vascular bundle, particularly of gymnosperms (see pp. 35 & 64)	transport of water	page 7
phloem	in the vascular bundle	transport of sugar, amino acids etc.	page 8
vascular cambium	in the vascular bundle	when it becomes active it forms cells that differentiate into secondary phloem, secondary xylem, and secondary fibres	page 9 page 16

Apical meristem

The **apical meristem** is a tiny region at the top of the stem. The outermost layers are called the *tunica*. In these layers of cells the divisions are all perpendicular to the surface (anticlinal) and the cells cut off differentiate into epidermis and also contribute to the developing leaves (leaf primordia). Below the tunica is the *corpus* which is far less homogeneous. Cell divisions occur in all directions, and the cells cut off differentiate into the cells of the *pith* (stem centre), cells of the *vascular bundles* (via procambial strand cells), and cells of the *cortex* (excluding the epidermis).

Thus active cell division of the stem apical meristem produces leaves in addition to stem tissues. Many plants have a single leaf at the level of each node (this arrangement is called *alternate)* and is illustrated in Figure 1.1(i) opposite, and in *Vicia faba* (Broad Bean). Some plants have leaves in pairs at each node, an arrangement called *opposite*. Each succeeding pair occurs at right angles to the previous, i.e. they are *decussate*. This decussate arrangement or any spiral arrangement prevents undue overshadowing.

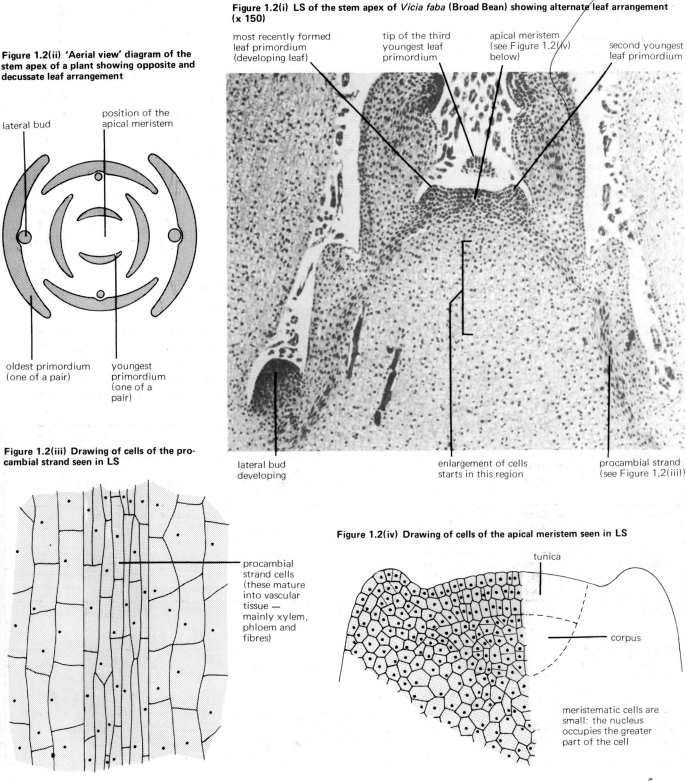

Figure 1.2(ii) 'Aerial view' diagram of the stem apex of a plant showing opposite and decussate leaf arrangement

lateral bud

position of the apical meristem

oldest primordium (one of a pair)

youngest primordium (one of a pair)

Figure 1.2(i) LS of the stem apex of *Vicia faba* (Broad Bean) showing alternate leaf arrangement (x 150)

most recently formed leaf primordium (developing leaf)

tip of the third youngest leaf primordium

apical meristem (see Figure 1.2(iv) below)

second youngest leaf primordium

lateral bud developing

enlargement of cells starts in this region

procambial strand (see Figure 1.2(iii))

Figure 1.2(iii) Drawing of cells of the procambial strand seen in LS

procambial strand cells (these mature into vascular tissue — mainly xylem, phloem and fibres)

Figure 1.2(iv) Drawing of cells of the apical meristem seen in LS

tunica

corpus

meristematic cells are small: the nucleus occupies the greater part of the cell

5

Figure 1.3(i) Annular vessels seen in LS (x 350) These vessels had been subject to stretching

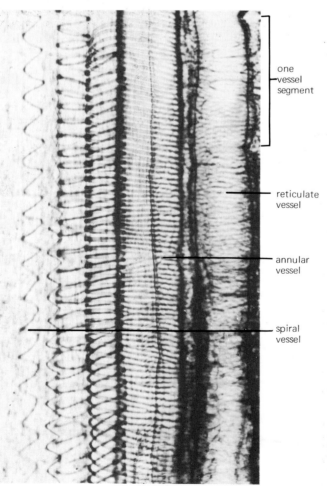

annular vessels have rings of thickening

Figure 1.3(ii) Drawing of stages in the formation of an annular vessel seen in LS

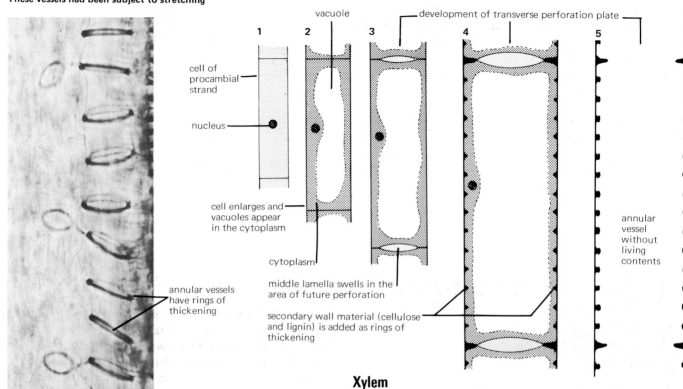

vacuole

development of transverse perforation plate

1 2 3 4 5

cell of procambial strand

nucleus

cell enlarges and vacuoles appear in the cytoplasm

cytoplasm

middle lamella swells in the area of future perforation

secondary wall material (cellulose and lignin) is added as rings of thickening

annular vessel without living contents

Figure 1.3(iii) Primary xylem vessels seen in LS (x 550)

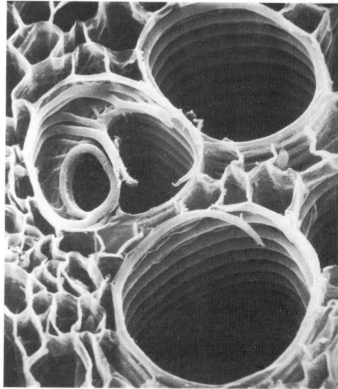

one vessel segment

reticulate vessel

annular vessel

spiral vessel

Xylem

Xylem vessels develop from procambial strand cells on the inner side of the strand. The first formed xylem vessels (called *proto-xylem*) consist of *annular vessels* (shown above), or vessels with spirally arranged thickening bands *(spiral vessels)*. In the growing stem the protoxylem differentiates and matures among actively elongating cells and is subject to stress. The vessels are stretched and many are eventually destroyed.

Figure 1.3(iv) Scanning electron micrograph of xylem vessels of *Nymphoides peltata* (petiole) (x12 000)

The later formed primary xylem is called *metaxylem*. It matures after elongation of the stem is completed. The walls are massively thickened with cellulose impregnated with lignin, but with irregular areas of unthickened wall *(reticulate vessel)* or with organized pit areas *(pitted vessel)*.

Upward movement of water occurs from the root hair region of the root tip (see page 41) to all the extremities of the aerial system. Water movement occurs in xylem vessels connected together end to end to form non-living tubes. Water is drawn up the stem by a force generated in the leaves when evaporation occurs from the mesophyll cells there (see page 29). Therefore the water column in xylem vessels is under tension rather than under pressure. The thickenings in xylem vessels occur on the inner surface of the walls and resist the tendency for the vessels to collapse under tension.

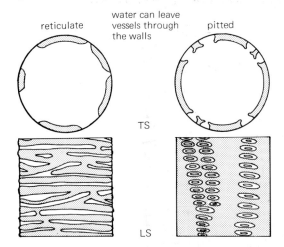

Figure 1.4(i) Drawings of metaxylem vessels seen in TS and LS

reticulate

water can leave vessels through the walls

pitted

TS

LS

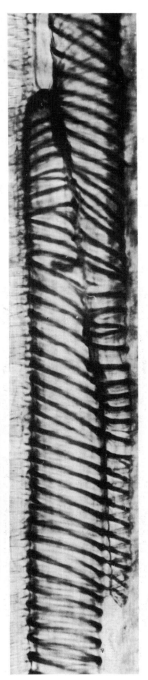

Figure 1.4(ii) Point of overlap of two tracheids seen in LS in the stem of *Cucurbita pepo* (Vegetable Marrow) (x 250)

movement of water through the pit membrane

Tracheids. In gymnosperms (and also in ferns) the xylem tissue consists of tracheids. Tracheids are long, narrow cells with pointed ends. They are non-living when mature, and have various forms of lignified thickening in their walls. They have no perforation plates. The passage of water from tracheid to tracheid occurs through pairs of pits in adjacent tracheids, but a thin primary wall separates each tracheid lumen (whereas in xylem vessels water moves freely through perforated end walls). Tracheids do occur in some dicotyledons. However, the tracheids of angiosperms do not show bordered pits *with* tori.

Figure 1.4(iii) Drawings of tracheids of *Pinus sylvestris* (Scots Pine): left, seen in LS showing the role of bordered pits in water movement; below, stereoscopic view

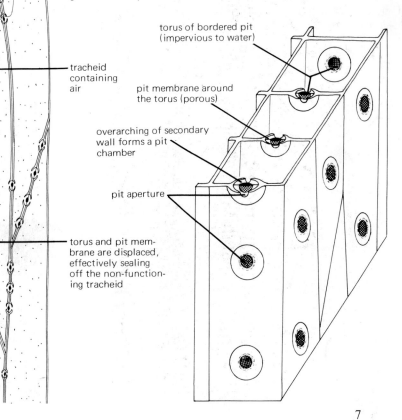

water is drawn up the stem by a force generated in the leaves due to evaporation

tracheid containing air

torus of bordered pit (impervious to water)

pit membrane around the torus (porous)

overarching of secondary wall forms a pit chamber

pit aperture

torus and pit membrane are displaced, effectively sealing off the non-functioning tracheid

Phloem

Phloem develops from procambial strand cells on the outer side of the strand. Phloem consists of *sieve tube* elements and *companion cells*. The first formed primary phloem (called *protophloem)* is often crushed in the development of the later formed phloem *(metaphloem)*. Mature phloem cells have thin cellulose walls.

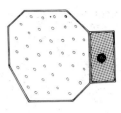

sieve tube and companion cell seen in TS above the sieve plate

Figure 1.5(i) Drawings of stages in the formation of a sieve tube and companion cell

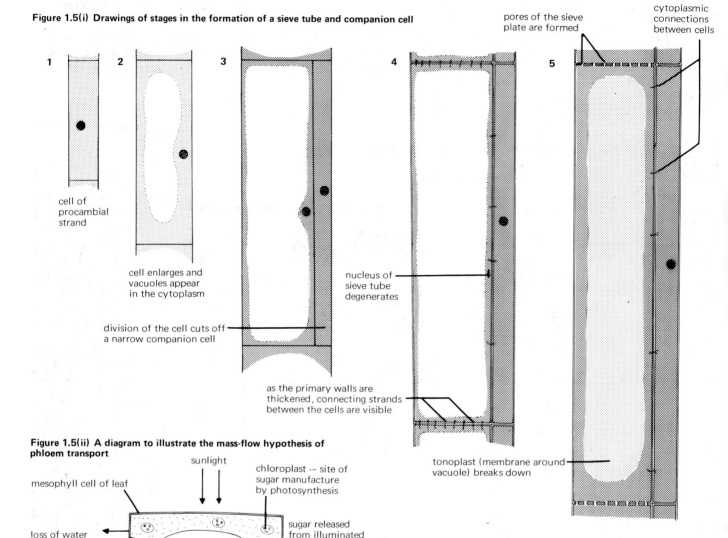

1 cell of procambial strand

2 cell enlarges and vacuoles appear in the cytoplasm

division of the cell cuts off a narrow companion cell

3 as the primary walls are thickened, connecting strands between the cells are visible

4 nucleus of sieve tube degenerates

tonoplast (membrane around vacuole) breaks down

pores of the sieve plate are formed

cytoplasmic connections between cells

5

Figure 1.5(ii) A diagram to illustrate the mass-flow hypothesis of phloem transport

sunlight

mesophyll cell of leaf

chloroplast -- site of sugar manufacture by photosynthesis

loss of water by evaporation

sugar released from illuminated chloroplast generates high osmotic potential here

cytoplasmic connections between cells (plasmodesmata)

high hydrostatic pressure in the mesophyll cell leads to bulk export along the sieve tubes

transpiration stream of water up the xylem vessels

phloem sieve tube and companion cell

region of low osmotic potential in starch storage cell, e.g. in root

uptake of water from soil

water returns to leaf in the transpiration stream

starch grain

Phloem sieve tubes are the site of solute transport in the plant (i.e. sugar transported as sucrose from the leaves to all parts of the plant, amino acids - particularly aspartic and glutamic acids and the associated amides - to all parts of the plant).

Transport in the phloem is a process involving living cells that are maintained with a supply of energy. Energy is involved in the 'loading' of solutes at their source, and 'unloading' of the sieve tubes at the 'sink' for that solute. Energy is required to maintain the living cells, and additional energy may be required to maintain movement along the tubes.

Several hypotheses have been suggested for the mechanism of phloem transport, but the subject remains one of controversy. For Munch's pressure flow hypothesis to operate there must be a gradient of osmotic potential along the sieve tube in the direction source ⟶ sink, capable of generating a turgor gradient to achieve bulk movement of water and solutes together. One alternative view is that streaming (cyclosis) of the cytoplasm in sieve tubes, and transcellular streaming of cytoplasm along the whole sieve tube, is responsible for the transport of solutes.

Figure 1.6(i) Primary phloem of *Cucurbita pepo* (Vegetable Marrow) seen in TS (x 850)

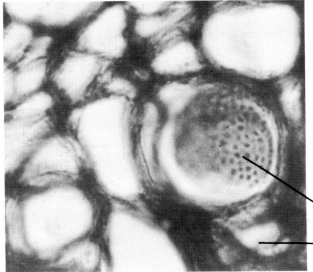

sieve plate showing pores

companion cell

Sclerenchyma

Sclerenchyma is thick-walled tissue that is relatively hard and rigid. Its principal function is support. The walls are composed of layers of cellulose impregnated with lignin. Pit connections between cells occur commonly, but the cell protoplast (cytoplasm and nucleus) often dies off at maturity of the cell.

Sclerenchyma exists as long, narrow, pointed cells called *fibres,* and as some shorter, irregularly shaped cells called *sclereids* (see page 26).

Fibres occur in and around the vascular tissue, and may also develop below the epidermis as a cylinder of supporting tissue in some older stems. Sclereids may occur singly or in groups in stems, leaves, fruits, and seeds.

Figure 1.6(ii) Primary phloem of *Cucurbita pepo* seen in LS (x 400)

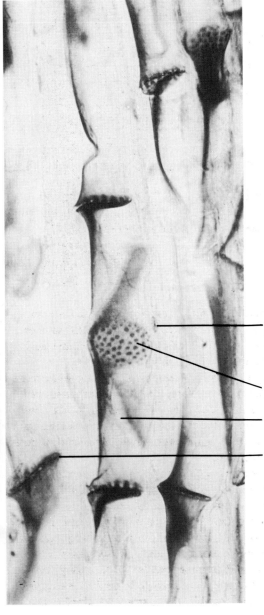

nucleus of companion cell

sieve plate in face view

sieve tube

sieve plate in section

Figure 1.6(iii) Fibres of *Zea mays* (Maize) stem seen in TS (x 400)

Figure 1.6(iv) Drawing of fibres seen in LS

middle lamella

simple pit connecting empty lumen of mature fibres (dead cells)

pointed ends of fibres interlock without air spaces

wall — layers of cellulose impregnated with lignin

Vascular cambium develops from the procambial strand cells remaining between the metaphloem and the metaxylem in the primary vascular bundle. The structure and role of cambium in the stem is shown on page 16.

9

Support, storage, protection

Parenchyma in the primary stem growth develops from the cells cut off by the corpus region of the meristem. Parenchyma is tissue which exhibits relatively little specialization, and which may be concerned with various physiological functions of the plant including storage of starch. Parenchyma cells retain the ability to divide when mature, and they play an important part in wound recovery. Large portions of the plant consist of parenchyma (i.e. all the pith of the stem, most of the cortex of stem and root, the mesophyll of stem and leaves (see page 29) and the fleshy part of fruits). Parenchyma may also occur in vascular bundles. Parenchyma cells may be polyhedral and isodiametric, or slightly elongated and cylindrical in shape. The walls remain thin and are formed of cellulose together with other substances, mainly polysaccharides. Air spaces occur between the cells. Turgid parenchyma cells are incompressible and exert pressure on the epidermis, thus creating support for herbaceous stems and for leaves.

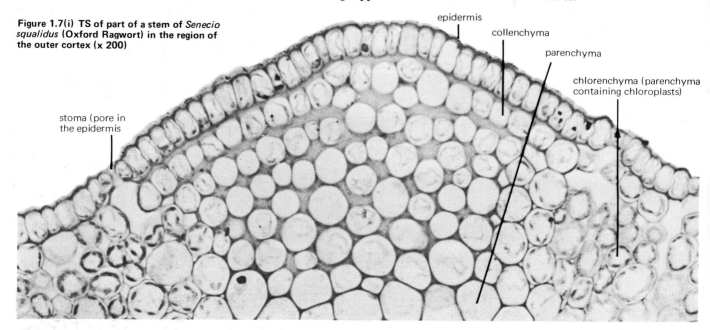

Figure 1.7(i) TS of part of a stem of *Senecio squalidus* (Oxford Ragwort) in the region of the outer cortex (x 200)

epidermis

collenchyma

parenchyma

chlorenchyma (parenchyma containing chloroplasts)

stoma (pore in the epidermis

Collenchyma consists of living cells, quite narrow and elongated in shape, with unevenly thickened cellulose walls. Collenchyma functions as a supporting tissue in young, growing stems, and it stretches with the growth of the stem or leaf in which it occurs. Collenchyma develops from cells cut off from the corpus region of the meristem, i.e. from the isodiametric cells of the meristem.
Epidermis is a continuous, compact layer of cells on the surface of the plant body. The layer is one cell thick and the cells are often tabular in shape. The epidermis of aerial parts of plants contains stomata (see pages 32 and 33), and all epidermal cells of the stem and the leaf have a waxy cuticle secreted on their outer walls. The epidermis often bears unicellular or multicellular hairs (see pages 37 and 38), or root hairs (see page 41). The epidermis is a thin skin which gives mechanical support and protection, and is concerned with the restriction of transpiration and the aeration of the plant body.

Figure 1.7(ii) Parenchyma cells of *Cucurbita pepo* in LS (x 250)

Figure 1.7(iii) Collenchyma cells in LS

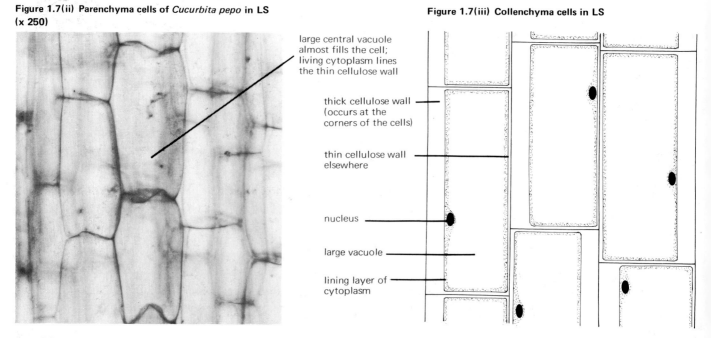

large central vacuole almost fills the cell; living cytoplasm lines the thin cellulose wall

thick cellulose wall (occurs at the corners of the cells)

thin cellulose wall elsewhere

nucleus

large vacuole

lining layer of cytoplasm

10

Figure 1.8(i) Stereogram of part of a stem of *Ranunculus acris* (Meadow Buttercup) to show how the tissue combats the stress forces experienced

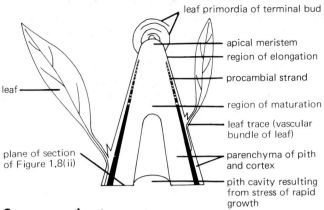

- leaf primordia of terminal bud
- apical meristem
- region of elongation
- procambial strand
- region of maturation
- leaf trace (vascular bundle of leaf)
- parenchyma of pith and cortex
- pith cavity resulting from stress of rapid growth
- leaf
- plane of section of Figure 1.8(ii)

Stresses on the stem

The stem may be compared with a modern ferro-concrete building. The hard, inextensible steel girders (vascular bundles) are surrounded by the softer incompressible concrete (turgid parenchyma). The load on the stem is the combined weight of leaves, flowers and fruits, and the stem resists compression. In the wind the aerial system offers resistance, and the windward side is under tension while the leeward side is under compression.

- vascular bundle
- parenchyma
- compression due to weight of plant
- wind
- compression due to wind
- tension due to wind

Figure 1.8(ii) TS of a mature stem of *Ranunculus acris* (x 20)

- cortex
- epidermis
- parenchyma
- vascular bundle
- leaf traces
- pith cavity

Dicotyledons and monocotyledons

The **angiosperms** (flowering plants) are divided into **dicotyledons** and **monocotyledons**, based on the number of 'seed leaves' present in the embryo. (Cotyledons are simpler than the normal leaf which develops later.) The common differences shown are:

Dicotyledons (e.g. Buttercup, Sunflower, Lime tree)	Monocotyledons (e.g. Iris, Maize, Wheat)
two seed leaves	one seed leaf (page 62)
usually have broad leaves with the veins forming a network (page 29)	have bayonet or strap-shaped leaves with parallel venation (page 36)
have flower parts in 2's (or multiples) or 5's	have flower parts in 3's (or multiples) (page 55)
have a small number of protoxylem groups in the root (page 43)	the number of protoxylem groups in the root is large (page 44)
vascular bundles of the stem in a ring (Figure 1.8(ii) and page 14)	vascular bundles more numerous and scattered throughout the stem (page 12)
have cambium in the mature vascular bundles (pages 16 and 48)	without cambium, and showing no secondary thickening

Figure 1.8(iii) TS of outer region of a stem of *Ranunculus acris* (x 200)

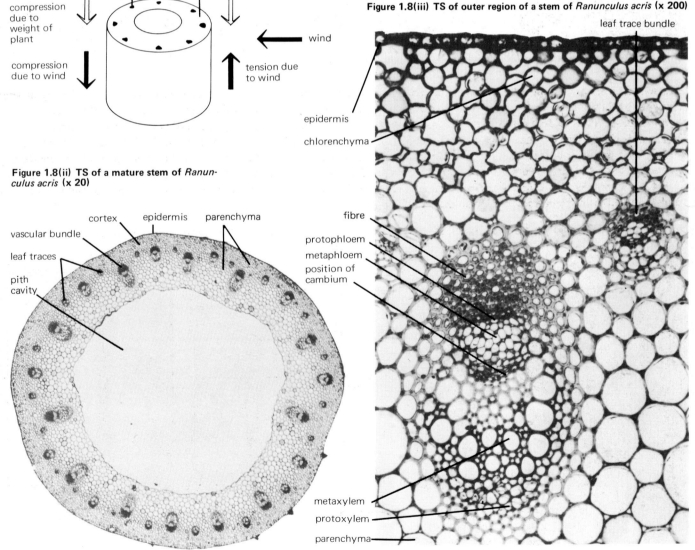

- leaf trace bundle
- epidermis
- chlorenchyma
- fibre
- protophloem
- metaphloem
- position of cambium
- metaxylem
- protoxylem
- parenchyma

11

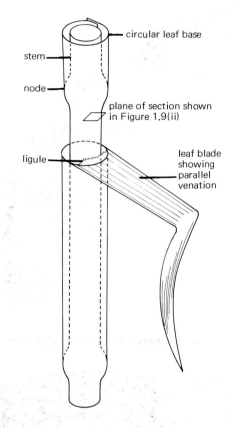

Figure 1.9(i) Drawing of part of a stem of *Zea mays* (Maize) showing the attachment of leaves

circular leaf base

stem

node

plane of section shown in Figure 1.9(ii)

ligule

leaf blade showing parallel venation

Figure 1.9(iii) Drawing of a longitudinal section through *Zea mays* stem showing how leaf trace bundles join with the outermost ring of vascular bundles

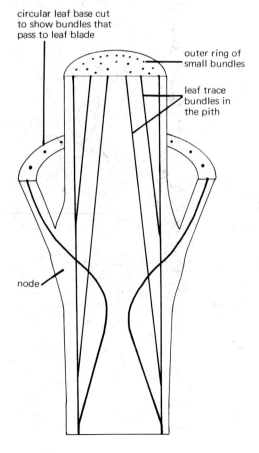

circular leaf base cut to show bundles that pass to leaf blade

outer ring of small bundles

leaf trace bundles in the pith

node

The stem in monocotyledons

Zea mays (**Maize**) is a monocotyledon. Together with Wheat (shown opposite) it is an example of a cultivated grass. Grasses are the most abundant and widely spread of all flowering plants. They provide our cereal grains, and the green herbage and dried fodder for our domestic animals. They provide us with fibres, sugar, oils, starch, protein, and alcohol. With the aid of their highly developed fibrous root system and vigorous leaf growth they cover the soil, and prevent erosion of soil by wind and water, especially on slopes.

The grass leaf has a blade that is characteristic of monocotyledons, but the leaf base is circular, surrounding the stem for some distance above the node. At the junction of base and blade is a small, membranous *ligule*.

Figure 1.9(ii) TS of part of a stem of *Zea mays* (x 60)

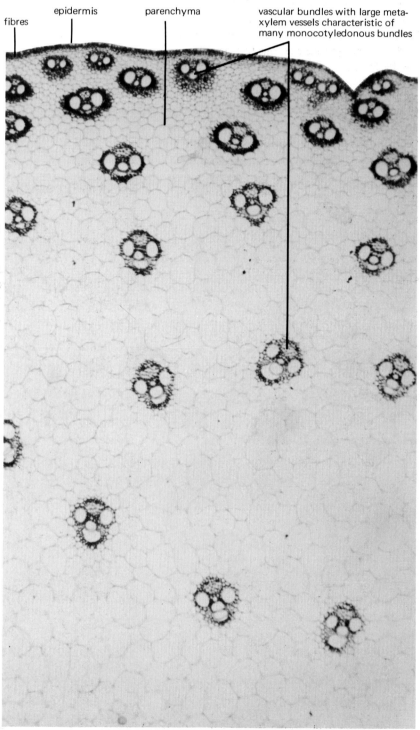

fibres

epidermis

parenchyma

vascular bundles with large meta-xylem vessels characteristic of many monocotyledonous bundles

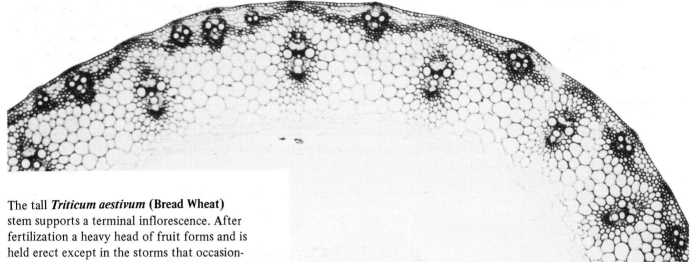

Figure 1.10(i) TS of part of a stem of *Triticum aestivum* (Bread Wheat) (x 10)

The tall **Triticum aestivum (Bread Wheat)** stem supports a terminal inflorescence. After fertilization a heavy head of fruit forms and is held erect except in the storms that occasionally flatten the harvest crops. Within the circular hollow stem the main bundles and *leaf trace* bundles form a compact girder system. The band of fibres below the epidermis gives additional strength.

Continued rapid elongation of the stem after the first xylem vessels have formed results in stretching of protoxylem to the point of rupture and destruction. All water transport then occurs in the large metaxylem vessels.

Figure 1.10(ii) TS of part of a stem of *Triticum aestivum* showing detail of the vascular bundle and leaf trace bundle (x 50)

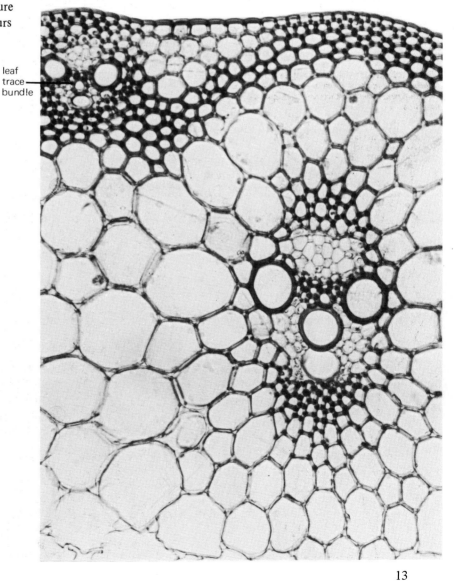

leaf trace bundle

Figure 1.10(iii) High power detail of a representative portion of the vascular bundle

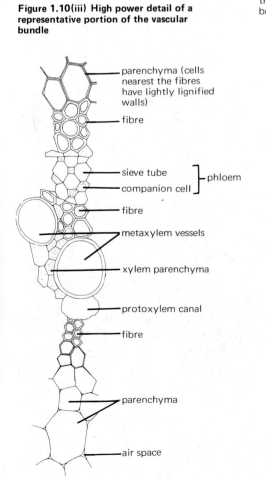

parenchyma (cells nearest the fibres have lightly lignified walls)

fibre

sieve tube — phloem
companion cell —

fibre

metaxylem vessels

xylem parenchyma

protoxylem canal

fibre

parenchyma

air space

The stem in dicotyledons

Cucurbita pepo (**Vegetable Marrow**) is a dicotyledonous plant which shows some unusual features. There are two rings of vascular bundles in the stem; the outer, smaller bundles alternate with the larger, inner bundles which border on the pith cavity. (A pith cavity arises in any stem in which the pith parenchyma cells have ceased growing whilst the whole stem continues to enlarge.)

An angled stem is less resistant to bending stress than a circular stem. *Cucurbita* stem has collenchyma below the epidermis, and a broad band of fibres in the cortex also. These layers provide strength to the pentagonal-shaped stem. In each vascular bundle there is phloem both internal and external to the xylem; this condition is described as *bicollateral*. It occurs in some families of flowering plants, such as the Cucurbitaceae (to which Vegetable Marrow belongs) and the Solanaceae (to which the Potato, Tomato, and Tobacco plants belong). The more common arrangement of phloem in the bundle is seen in the Sunflower plant (shown opposite) and is called *collateral*.

Figure 1.11(i) TS of a stem of *Cucurbita pepo* (Vegetable Marrow) (x 10)

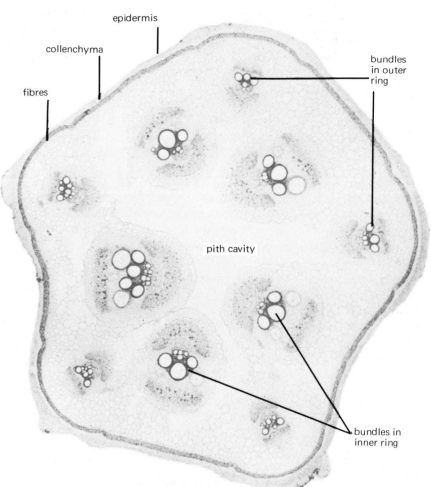

Figure 1.11(ii) TS of the cortex of *Cucurbita pepo* showing a vascular bundle in detail (x 60)

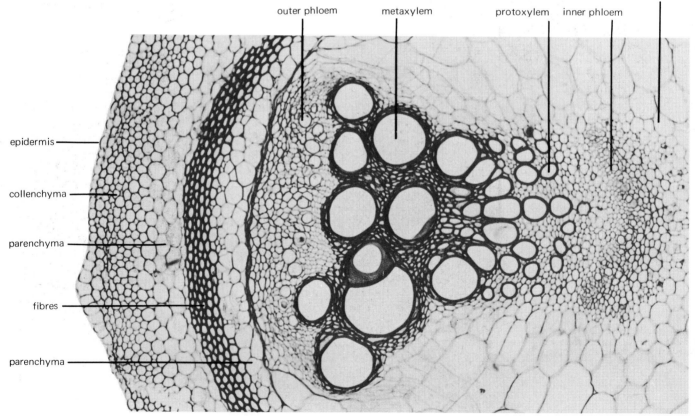

14

Figure 1.12(i) TS of part of a stem of *Helianthus annuus* (Sunflower) (x 30)

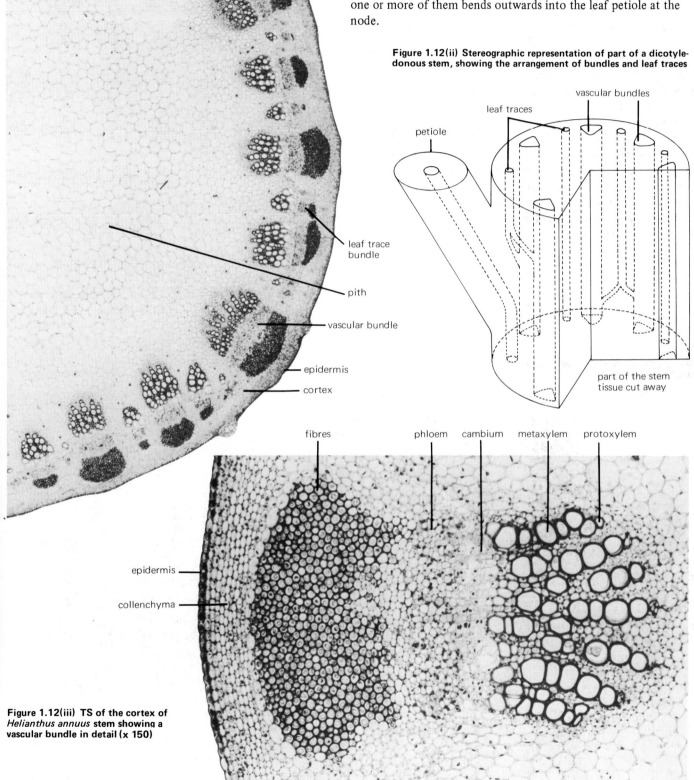

leaf trace bundle

pith

vascular bundle

epidermis

cortex

Helianthus annuus (**Sunflower**) belongs to the Compositae family. This is the largest family of flowering plants, comprising 900 genera and 13 000 species. Species are found in almost every habitat. The inflorescence is characteristic, being a *capitulum* (see page 55). In *Helianthus annuus* the stem may be 2–2.5 m high, and supports a massive capitulum more than 30 cm broad. The bundles and leaf traces, each with a massive cap of fibres, form a girder system. As in most dicotyledonous stems, leaf trace bundles branch from the main bundles well below the nodes. These leaf traces occur between the main bundles, and one or more of them bends outwards into the leaf petiole at the node.

Figure 1.12(ii) Stereographic representation of part of a dicotyledonous stem, showing the arrangement of bundles and leaf traces

leaf traces

vascular bundles

petiole

part of the stem tissue cut away

fibres   phloem   cambium   metaxylem   protoxylem

epidermis

collenchyma

Figure 1.12(iii) TS of the cortex of *Helianthus annuus* stem showing a vascular bundle in detail (x 150)

15

The older stem–secondary thickening

Secondary thickening occurs in shrubs and trees and also in many herbaceous plants. An example of the latter is *Helianthus annuus,* shown here. Secondary thickening results in additional xylem and phloem tissue being formed, in rows that are more regular when compared with primary growth. The pith tissue is unchanged as a result of secondary thickening, but the cortex is progressively stretched as new cells are formed and the stem increases in girth. Secondary tissue facilitates transport of water and solutes in the plant and greatly enhances the strength of the stem.

The first step in secondary thickening involves the cambium of the bundles *(fascicular cambium)* being connected up by the formation of *interfascicular* cambium in the parenchyma between the bundles. A cylinder of meristematic cells is formed in the stem. Then rows of secondary xylem vessels and fibres are formed towards the centre of the stem, and rows of secondary sieve tubes with companion cells are formed towards the outside.

Figure 1.13(i) TS of the outer part of a stem of *Helianthus annuus* (Sunflower) showing an early stage in secondary thickening (x 80)

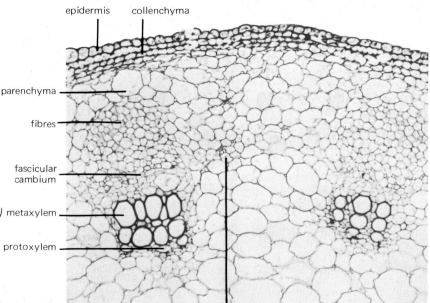

epidermis collenchyma

parenchyma

fibres

fascicular cambium

metaxylem

protoxylem

interfascicular cambium developed in parenchyma cells

Figure 1.13(ii) Stages in the formation of secondary tissue

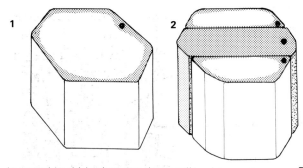

In secondary thickening parenchyma cells between the bundles become meristematic

Cambium cells cut off new cells both to the inside and the outside of the cambium ring

Outer cells enlarge and differentiate into secondary phloem tissue

Inner cells enlarge and differentiate into secondary xylem tissue devoid of contents

Figure 1.13(iii) TS of part of a stem of *Helianthus annuus* showing primary and secondary vascular tissue (x 80)

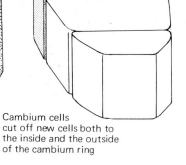

cortex, stretched

fibres

primary phloem

secondary phloem

cambium

secondary xylem

primary xylem (meta- and protoxylem)

16

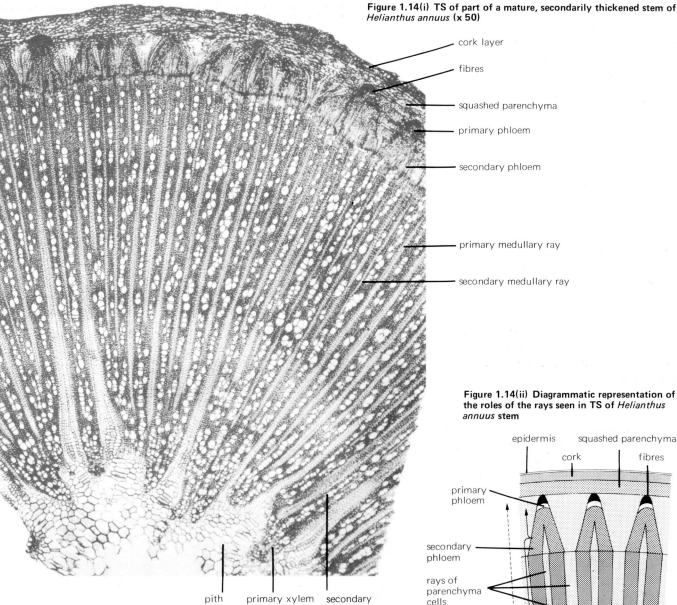

Figure 1.14(i) TS of part of a mature, secondarily thickened stem of *Helianthus annuus* **(x 50)**

cork layer

fibres

squashed parenchyma

primary phloem

secondary phloem

primary medullary ray

secondary medullary ray

pith   primary xylem   secondary xylem

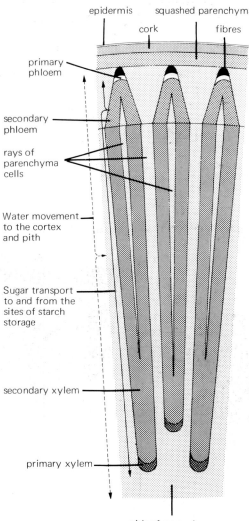

Figure 1.14(ii) Diagrammatic representation of the roles of the rays seen in TS of *Helianthus annuus* **stem**

epidermis   squashed parenchyma

cork   fibres

primary phloem

secondary phloem

rays of parenchyma cells

Water movement to the cortex and pith

Sugar transport to and from the sites of starch storage

secondary xylem

primary xylem

pith of parenchyma

The ring of cambium cuts off more cells towards the inside to differentiate into xylem tissue than it does to the outside to become phloem. Phloem is a delicate, thin-walled tissue and the older phloem collapses and becomes compressed as it ages, whereas the xylem tissue is strengthened with cellulose and lignin and remains uncrushed. Consequently secondary xylem cells appear more numerous than phloem cells.

The steadily increasing girth of the cambium ring and of the outer part of the stem is catered for by frequent *radial divisions* in the cambium that produce more cambium cells. This is in addition to the *tangential divisions* that produce xylem and phloem cells from the cambium. The pressure due to the presence of additional secondary xylem and phloem squashes the cortex. Splitting of the cortex is prevented by a *cork cambium* developing in the cortex (page 22). Columns of parenchyma cells, called *medullary rays*, connect the living cells of the cortex with the living cells of the pith. The rays facilitate the transport of sugars, water, and oxygen between vascular tissue and the parenchymatous sites of starch storage. Rays in existence at the beginning of secondary thickening (i.e. parenchyma between the primary vascular bundles) are called *primary rays* and connect the cortex with the pith. Later-formed rays (i.e. *secondary rays*) are shorter. They develop from *ray initials* (see page 19) that are small cambium cells formed from radial divisions of the cambium ring.

The older stem–wood

Woody plants show differences in their secondary thickening from that in annual plants like *Helianthus annuus* (Sunflower).

The cambium of woody plants develops as a *completed ring* in the primary stem, and the initial production of primary xylem and phloem is minimal. Virtually all the vascular tissue in a tree is secondary tissue. The stem of woody plants persists from year to year and additional secondary xylem and phloem are formed each year. In temperate climates growth takes place in spring and summer. (In plants of warm climates growth is also periodic, ceasing in the dry season, or during a resting period.) The first-formed xylem each season contains large, thin-walled vessels and fibres and is called *early wood*. Towards the end of the growth period the vessels and fibres are small and thick-walled, and this *late wood* has a dense, dark appearance and gives the annual growth ring. These annual rings are really transverse sections through growth cones. During each succeeding growth season the woody stem increases in height by means of its apical meristem, and in girth by the activity of its vascular cambium. The stem or trunk of a tree is a *very elongated cone,* not a cylinder.

In branches and in leaning trunks the pith may be eccentric with much wider growth rings on one side than on the other. A branch is like a cantilever beam in that it is fixed at one end only. The weight of the branch causes the lower surface to be in compression and the upper surface to be under tension. In transverse section such branches show *reaction wood,* and in all hardwood varieties (i.e. dicotyledonous trees) this is formed on the side of the branch under tension.

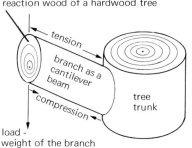

reaction wood of a hardwood tree

tension

branch as a cantilever beam

compression

tree trunk

load - weight of the branch

Figure 1.15(i) TS of a stem of *Tilia vulgaris* (Lime) one year old (x 12)

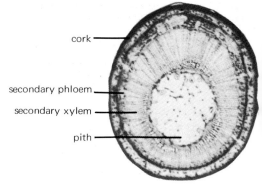

cork

secondary phloem

secondary xylem

pith

Figure 1.15(ii) Part of a TS of a four-year-old twig of *Tilia vulgaris* (x 30)

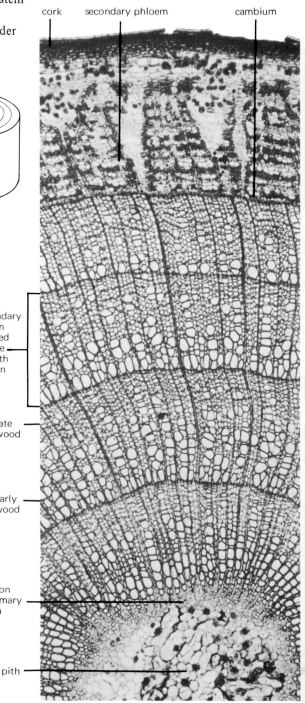

cork   secondary phloem   cambium

secondary xylem formed in one growth season

late wood

early wood

position of primary xylem

pith

Figure 1.15(iii) TS of a six-year-old twig of *Tilia vulgaris* showing reaction wood (x 10)

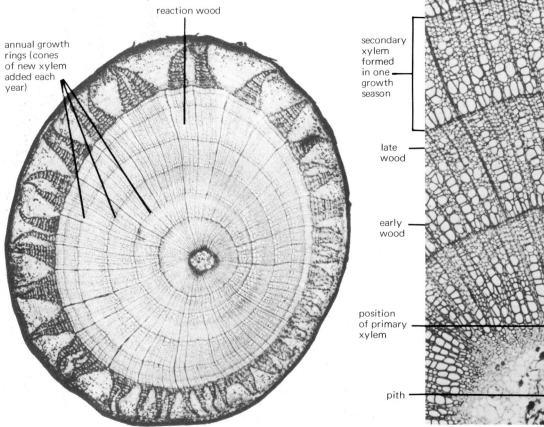

reaction wood

annual growth rings (cones of new xylem added each year)

Formation of rays in wood

Figure 1.16(i) TS of part of a stem of *Fraxinus excelsior* (Ash) in the region of secondary xylem (x 150)

Figure 1.16(ii) Drawings of stages in the formation of a ray initial in the cambium. 1—3 are stages in cambial activity seen in TS; right, stereoscopic view of ray initial and cambium cells

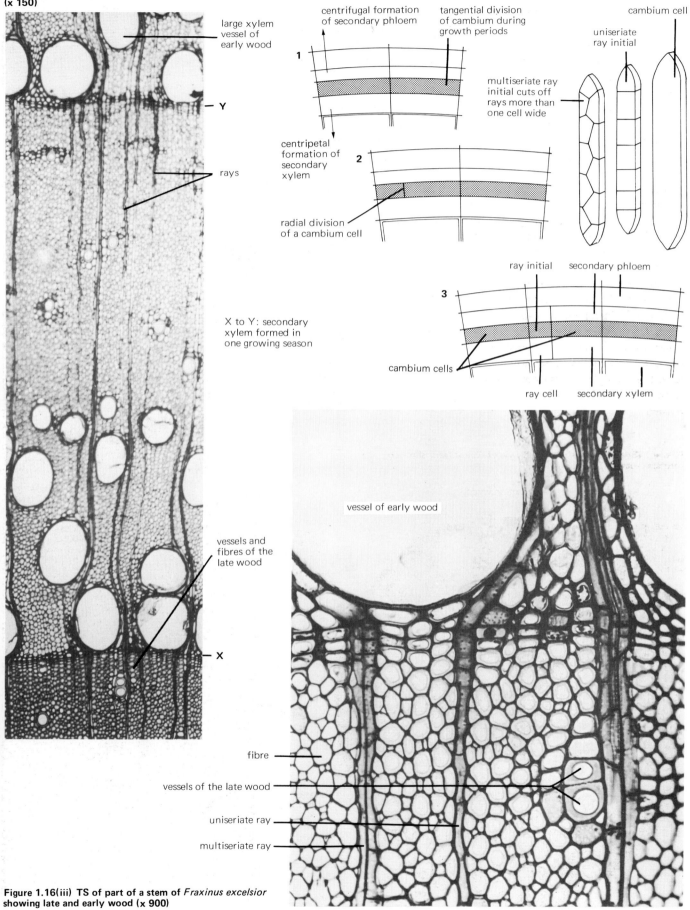

large xylem vessel of early wood

— Y

rays

centrifugal formation of secondary phloem

tangential division of cambium during growth periods

cambium cell

uniseriate ray initial

multiseriate ray initial cuts off rays more than one cell wide

1

centripetal formation of secondary xylem

2

radial division of a cambium cell

X to Y: secondary xylem formed in one growing season

ray initial   secondary phloem

3

cambium cells

ray cell   secondary xylem

vessels and fibres of the late wood

— X

vessel of early wood

fibre

vessels of the late wood

uniseriate ray

multiseriate ray

Figure 1.16(iii) TS of part of a stem of *Fraxinus excelsior* showing late and early wood (x 900)

19

Pits and tyloses

When the cell wall is formed between two new cells the cytoplasm of the cells remains connected. Very thin protoplasmic connections *(plasmodesmata)* pass through the *middle lamella*, and as the primary wall is built up this region remains thin and becomes a primary **pit** field.

When secondary wall layers are added the pits become more easily visible as extremely thin areas in the walls. Such pits may be *simple*, or the secondary wall may partially arch over the pit area to form a *bordered* pit.

Figure 1.17(i) Drawing of part of a twig cut to show the planes of radial longitudinal section (RLS) and tangential longitudinal section (TLS)

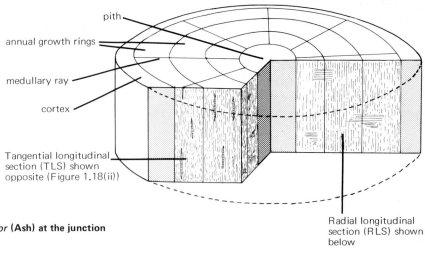

pith

annual growth rings

medullary ray

cortex

Tangential longitudinal section (TLS) shown opposite (Figure 1.18(ii))

Radial longitudinal section (RLS) shown below

Figure 1.17(ii) RLS of part of a stem of *Fraxinus excelsior* (Ash) at the junction of late and early wood (x 300)

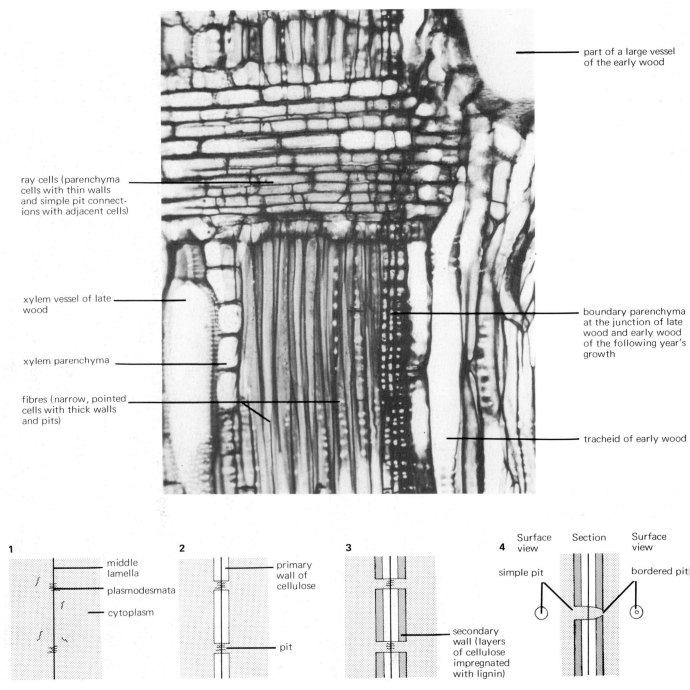

ray cells (parenchyma cells with thin walls and simple pit connections with adjacent cells)

xylem vessel of late wood

xylem parenchyma

fibres (narrow, pointed cells with thick walls and pits)

part of a large vessel of the early wood

boundary parenchyma at the junction of late wood and early wood of the following year's growth

tracheid of early wood

1

middle lamella

plasmodesmata

cytoplasm

2

primary wall of cellulose

pit

3

secondary wall (layers of cellulose impregnated with lignin)

4 Surface view   Section   Surface view

simple pit

bordered pit

Tyloses are outgrowths from a ray cell or xylem parenchyma cell through a pit in the wall into a xylem vessel or tracheid, partially or completely blocking the lumen. This may occur when water transport ceases, or after vessels have been injured or have become diseased. A tylosis is therefore an extension of a living cell, containing protoplasm, which secretes a wall within the vessel.

Figure 1.18(i) Stereogram of a cube of secondary xylem showing a ray in section

secondary xylem cells are formed in regular rows

longitudinal walls thickened with cellulose and lignin

transverse walls are dissolved away to provide a long continuous tube through which water is drawn with minimum resistance

Figure 1.18(ii) Tyloses in a xylem vessel

- parenchyma cell
- vessel wall
- tylosis
- vessel lumen

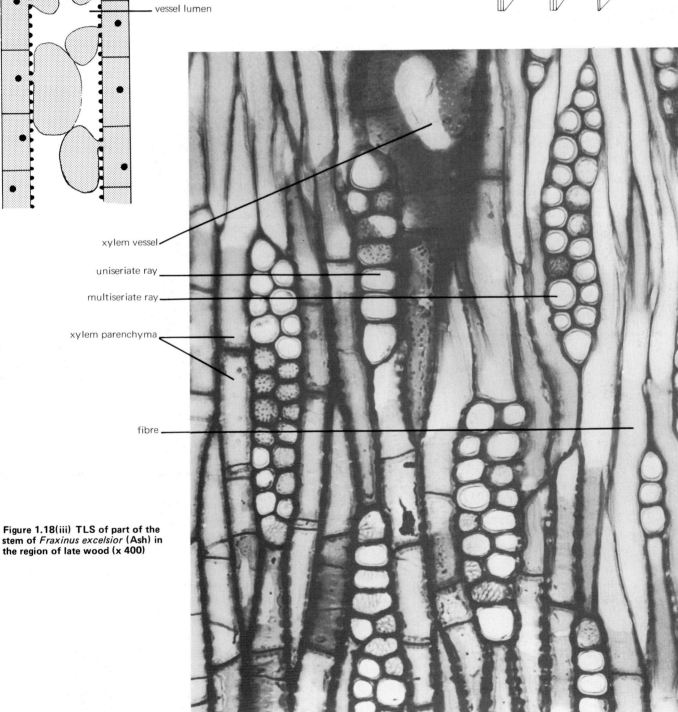

- xylem vessel
- uniseriate ray
- multiseriate ray
- xylem parenchyma
- fibre

Figure 1.18(iii) TLS of part of the stem of *Fraxinus excelsior* (Ash) in the region of late wood (x 400)

Cork and lenticels

After secondary thickening has begun in a twig or stem the epidermis and cortex are gradually replaced by the formation of **cork**. Cork cells are dead cells of a regular shape, without air-spaces in between them and with walls lined internally with *suberin* (wax-like material). The suberized walls are impervious to water and gases, and resistant to mechanical damage and attack by micro-organisms. The cork cambium *(phellogen)* arises by tangential cell division, usually in the layer of cells immediately below the epidermis (as in *Syringa vulgaris* (Lilac), shown opposite, although in *Ribes nigrum* (Blackcurrant) shrub, shown below, it arises in the primary phloem. The phellogen cuts off cork cells to the outside (i.e. centrifugally) and cuts off one or more layers of parenchyma cells (called the *phelloderm)* to the inside (centripetally).

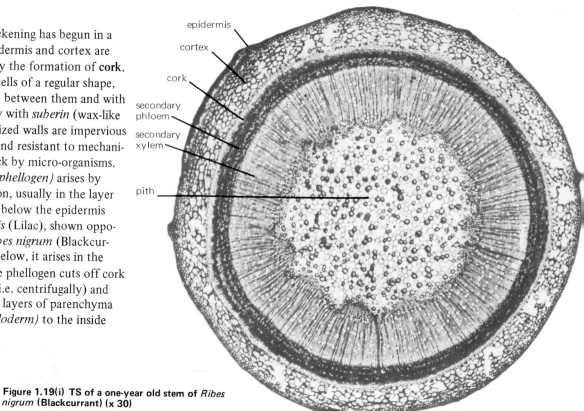

epidermis
cortex
cork
secondary phloem
secondary xylem
pith

Figure 1.19(i) TS of a one-year old stem of *Ribes nigrum* **(Blackcurrant) (x 30)**

Figure 1.19(ii) TS of part of the cortex of *Ribes nigrum* **showing cells formed by the phellogen (x 900)**

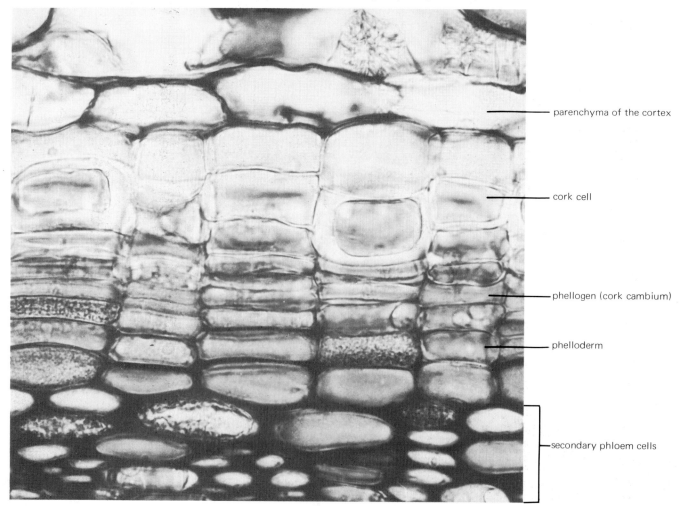

parenchyma of the cortex

cork cell

phellogen (cork cambium)

phelloderm

secondary phloem cells

The **lenticel** is a small area in the bark where the phellogen is more active, and divides to form loosely arranged cells in place of cork cells. The lenticel contains large and rounded cells that are usually unsuberized and are called complementary cells. The air spaces of the lenticel are continuous with the air spaces of the cortex, and it is assumed that the function of lenticels is to facilitate gas exchange. In some plants the phellogen of the lenticel produces 'closing layers' of suberized cells that alternate with layers of the parenchyma cells and hold them together.

Figure 1.20(i) TS of a stem of *Syringa vulgaris* (Lilac) showing the random position of lenticels (x 30)

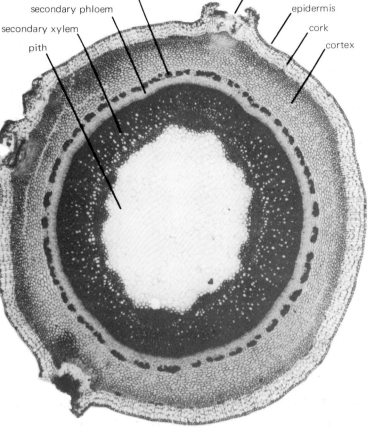

Figure 1.20(ii) Twig of *Syringa vulgaris* showing lenticels in the bark (x 8)

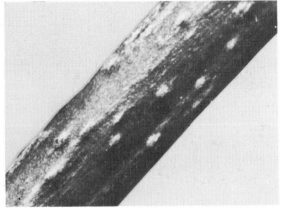

Figure 1.20(iii) TS of a lenticel in the stem of *Syringa vulgaris* (x 270)

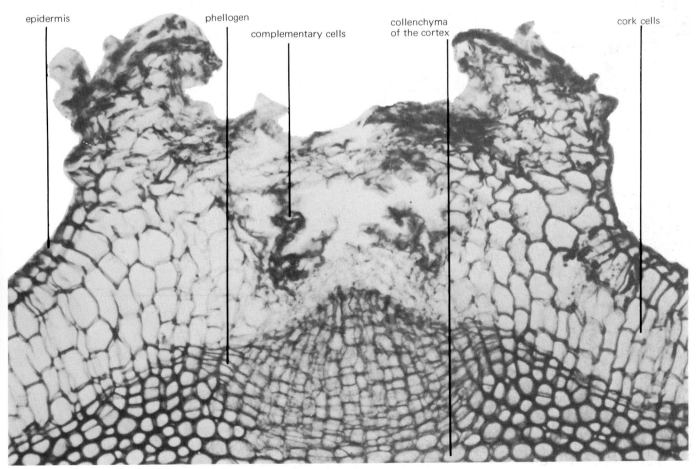

23

Photosynthetic stems

Juncus effusus (Soft Rush) is a locally dominant species in wet pastures, bogs, and damp woods, especially on acid soils. The leaves of this plant occur only as brown sheaths below the inflorescence, and the stem has a photosynthetic function. The epidermis has a thick outer wall and cuticle, and the stomata are sunken. Palisade mesophyll cells (see page 29) occur around the cortex except at the ridges which are supported by groups of fibres. Below the photosynthetic cortex are scattered vascular bundles typical of a monocotyledonous stem, and also large air spaces. The pith consists of stellate parenchyma.

Figure 1.21(i) *Juncus effusus* **(Soft Rush) growing in wet pasture**

Figure 1.21(ii) TS of part of a stem of *Juncus effusus* (x 80)

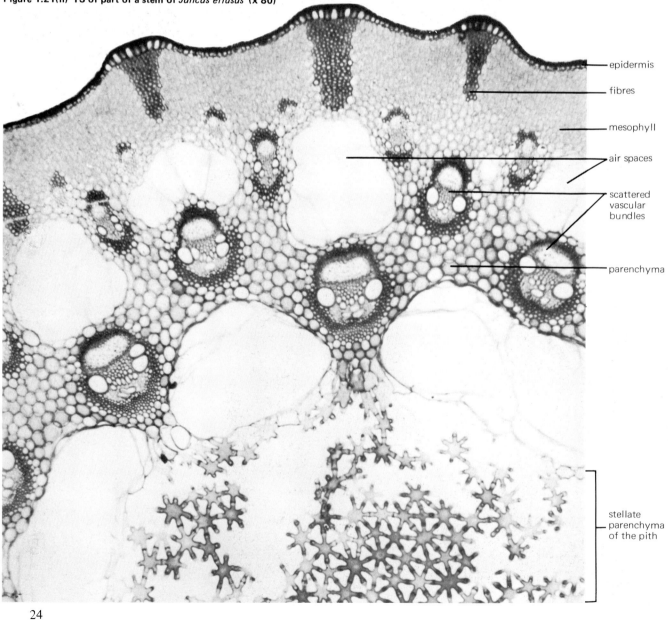

epidermis

fibres

mesophyll

air spaces

scattered vascular bundles

parenchyma

stellate parenchyma of the pith

24

In *Casuarina equisetifolia* (She Oak, an Australian tree) the leaves are scale-like, and the stem is the site of photosynthesis. Stomata occur only in the epidermis lining the sides of longitudinal grooves of the stem. Epidermal hairs occur in the grooves and assist in the trapping of moist air there. This reduces water loss from the stem. Palisade mesophyll cells occur between the grooves.

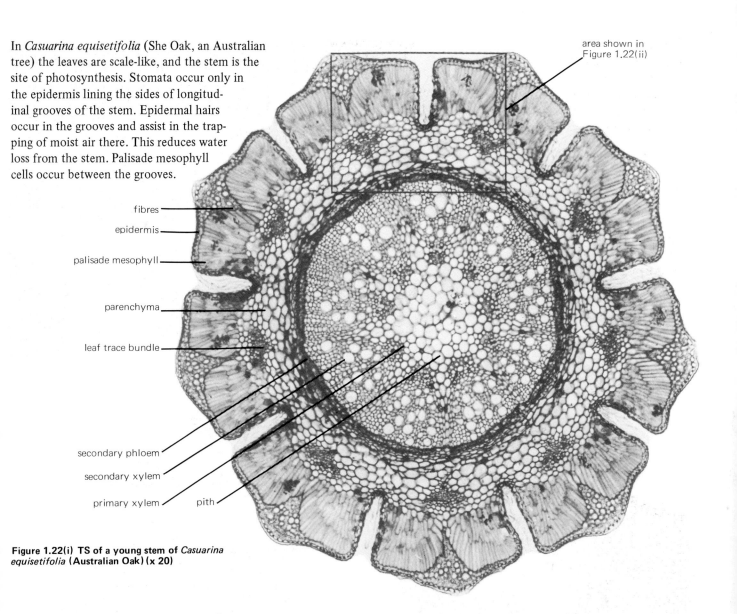

area shown in Figure 1.22(ii)

fibres

epidermis

palisade mesophyll

parenchyma

leaf trace bundle

secondary phloem

secondary xylem

primary xylem   pith

Figure 1.22(i) TS of a young stem of *Casuarina equisetifolia* (Australian Oak) (x 20)

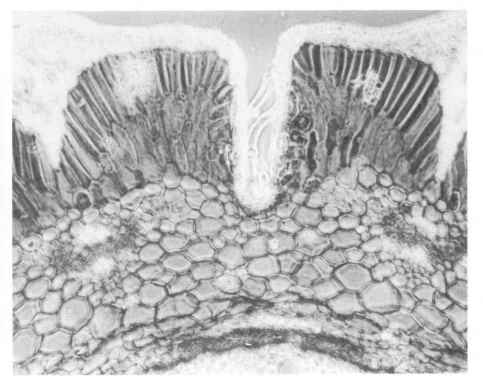

Figure 1.22(ii) TS of part of the stem of *Casuarina equisetifolia* (see Figure 1.22(i) for the photosynthetic and vascular tissue) (phase contrast, x 50)

25

Figure 1.23(i) VS of the midrib region of a floating leaf of *Nymphaea alba* (Water Lily) (x 110)

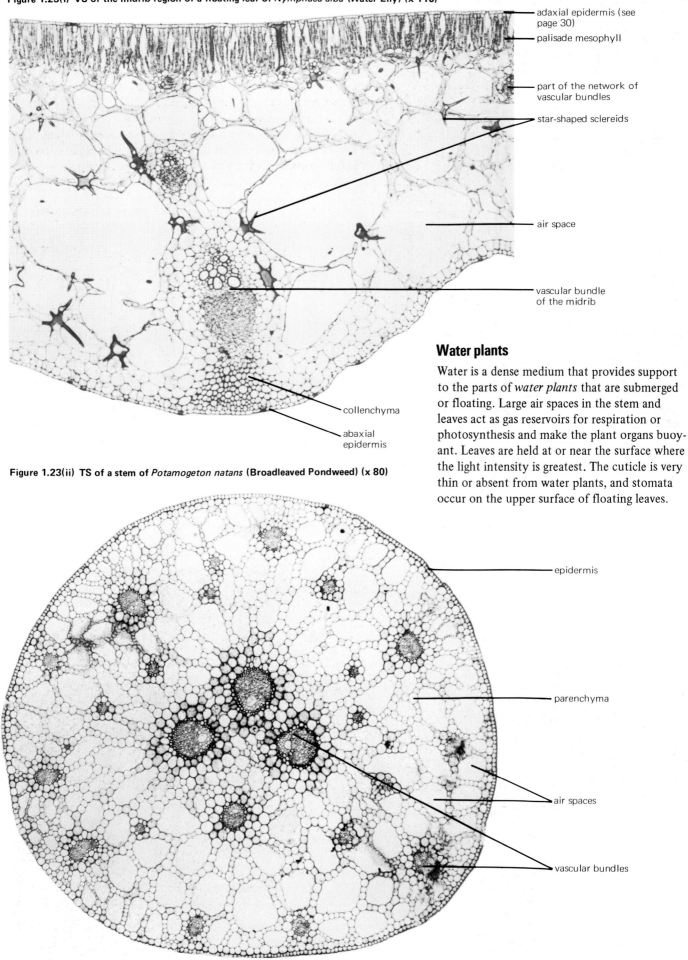

adaxial epidermis (see page 30)

palisade mesophyll

part of the network of vascular bundles

star-shaped sclereids

air space

vascular bundle of the midrib

collenchyma

abaxial epidermis

Figure 1.23(ii) TS of a stem of *Potamogeton natans* (Broadleaved Pondweed) (x 80)

epidermis

parenchyma

air spaces

vascular bundles

Water plants

Water is a dense medium that provides support to the parts of *water plants* that are submerged or floating. Large air spaces in the stem and leaves act as gas reservoirs for respiration or photosynthesis and make the plant organs buoyant. Leaves are held at or near the surface where the light intensity is greatest. The cuticle is very thin or absent from water plants, and stomata occur on the upper surface of floating leaves.

26

Figure 1.24(i) Host plant parasitized by *Cuscuta europaea* (Greater Dodder)

Figure 1.24(i) Host plant parasitized by *Cuscuta europaea* (Greater Dodder)

A parasitic stem

Cuscuta europaea (Dodder) is one of the few flowering plants that is completely parasitic. The seedling at germination has a tiny root and a long, thread-like stem that grows so that the tip shows growth movements in ever-widening circles *(nutates)*. If it meets a host plant such as clover, nettle, hop, gorse, or heather the stem twines around the host. Haustoria penetrate the host and the xylem and phloem of parasite and host are connected. The *Cuscuta* root then dies, the stem bears scale-like leaves and numerous small clusters of flowers. The *Cuscuta* plant contains no chlorophyll, and if it fails to make contact with a host plant it quickly dies.

Figure 1.24(ii) Tissue map of the connection between *Cuscuta europaea* and host stem shown in Figure 1.24(iii)

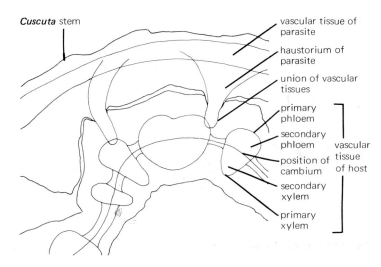

Cuscuta stem

vascular tissue of parasite

haustorium of parasite

union of vascular tissues

primary phloem

secondary phloem

position of cambium

secondary xylem

primary xylem

vascular tissue of host

Figure 1.24(iii) TS of the stem of a host plant and an oblique LS of the stem of *Cuscuta europaea* (x 30)

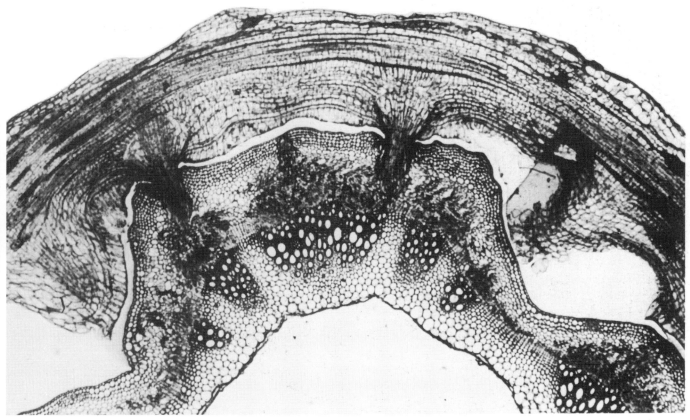

27

Section 2: The leaf-the factory for photosynthesis

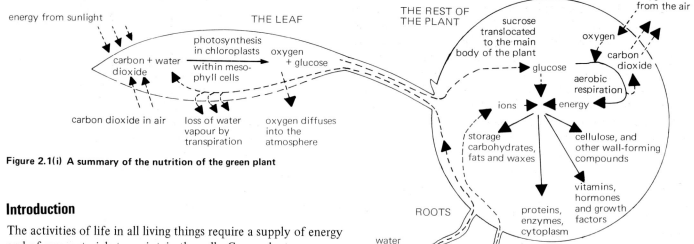

Figure 2.1(i) A summary of the nutrition of the green plant

Introduction

The activities of life in all living things require a supply of energy and of raw materials to maintain the cells. Green plants use energy from sunlight and the simple substances carbon dioxide and water from the environment to make glucose by a process called photosynthesis. This occurs mainly in the **leaf**. Glucose is then used by the plant as a whole to supply the energy and much of the materials to maintain the cells. In their nutrition green plants are thus directly independent of other living things.

Some idea of the extent of photosynthesis is gained from examining the carbon cycle in nature. All chemical elements (e.g. nitrogen, carbon) exist in finite quantities used in life, released in death and decay, and made available for re-use again and again. In the air the concentration of carbon dioxide remains low and fairly constant despite all the respiration, combustion, and decay that releases carbon dioxide into the air. Photosynthesis is the chief process that removes carbon dioxide from the air.

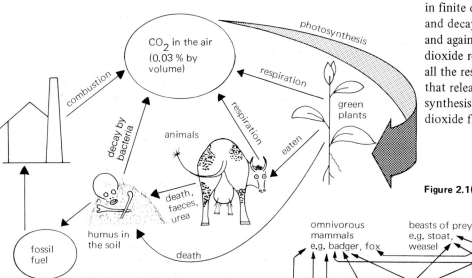

Figure 2.1(ii) Diagrammatic representation of the carbon cycle

Figure 2.1(iii) Simplified food web

Green plants are the primary producers of the community. They represent the broad base of a pyramid of numbers of organisms or of biomass in which a few tertiary consumers represent the pyramid top. All animal life depends directly or indirectly on green plants for energy-rich complex organic molecules, and for the maintenance of the composition of the air as we know it. All life ultimately depends on photosynthesis in plant leaves, with the exception of chemosynthetic organisms.

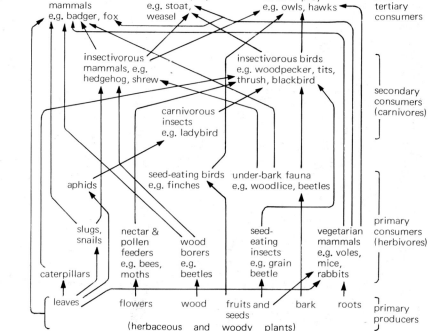

28

The leaf consists of a *lamina* or leaf blade connected to the stem by a *petiole* or leaf stalk. The lamina is a thin structure in which many cells are in positions receiving maximum sunlight. The whole leaf is supported by a system of branching veins that form a fine network throughout the lamina. A vein is a vascular bundle surrounded by a few parenchyma cells without chloroplasts (bundle sheath). Thus veins often appear lighter in colour than the rest of the leaf. The tough, continuous epidermis also provides support by counteracting the hydrostatic pressure of all the turgid cells of the leaf since it binds the cells together. The epidermis contains pores (called *stomata*) which facilitate gaseous exchange. Light is absorbed by chlorophyll pigments in the chloroplasts of the mesophyll cells. *Mesophyll* is the parenchymatous tissue internal to the transparent epidermis. The *palisade mesophyll* cells contain most chloroplasts. They receive water from the xylem vessels of the vein network. Carbon dioxide diffuses into the surface film of water around the mesophyll cells. Water inevitably evaporates from all internal surfaces and water vapour diffuses out of the stomata simultaneously. This is called *transpiration*. The sugar formed in photosynthesis is transported to the sieve tubes of the vein network and is *translocated* away.

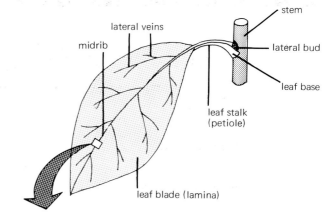

Figure 2.2(i) Diagram of a simple leaf

Figure 2.2(ii) Stereogram of part of the lamina of a leaf

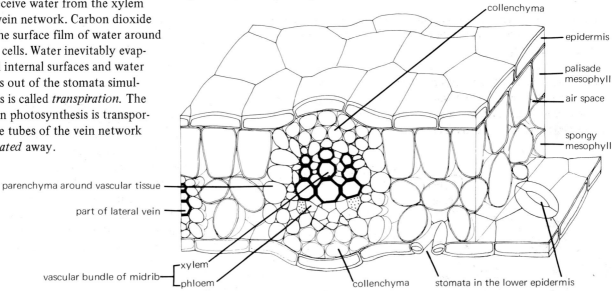

Figure 2.2(iii) Vein network of a dead leaf of *Buxus sempervirens* (Box) (x 100)

Figure 2.2(iv) High power detail of the lamina to show the site of photosynthesis in relation to the supply of raw materials

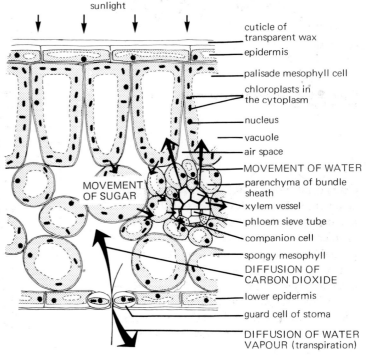

Deciduous leaves

The Plum tree *(Prunus domestica)* has deciduous leaves, as do most plants of temperate regions, where leaves are lost in the unfavourable season and new leaves grow the following spring. Most leaves grow and develop more or less at right angles to the light from the sky, and the upper part of the leaf which receives more light is quite different from the lower part. The leaf of *Prunus domestica* is differentiated in this way; such leaves are described as *dorsiventral.* The lamina of *Prunus domestica* leaf is mostly less than 1 mm thick (about 10–12 cells thick in fact), apart from the midrib area, yet the whole mature leaf is large, probably 7–12 cm long and 4–7 cm broad. Consequently, although the leaf is light, flexible, and supported internally, the leaf may be in danger of being detached from the stem because of its shape. The leaf base is very much wider than the rest of the petiole and is of a shape that helps resist the shearing forces generated in heavy rain or strong winds.

Figure 2.3(i) Leaf of *Prunus domestica* (Plum) with part of the lamina cut away, and showing the connection point of this leaf to the woody twig

lateral bud

position of leaf attachment to twig (node)

components of the shearing force that the leaf base may experience

petiole, flexibly supports the lamina in a position largely dictated by the direction of incident light

leaf base, forms a broad attachment with the stem and provides partial protection to the lateral bud in its axil

lamina with serrated margin

lateral veins

midrib, almost circular in cross section and containing a massive crescent-shaped vascular bundle

Figure 2.3(ii) VS of the lamina of a leaf of *Prunus domestica* (x 20)

adaxial (upper) epidermis

vascular bundle of midrib

spongy mesophyll

palisade mesophyll

small bundles of the lamina network

collenchyma at the leaf margin

parenchyma

abaxial (lower) epidermis

xylem tissue of rows of vessels surrounded by smaller fibres; the cell walls are thick, and strengthened with lignin

collenchyma

palisade mesophyll

spongy mesophyll

phloem tissue

parenchyma

Figure 2.3(iii) Tissue map of the lamina section in Figure 2.3(ii)

some cells contain resins, or pigments (tannins in the cytoplasm or anthocyanins in the vacuole); often the amount of these increases in the autumn

Figure 2.4(i) High power detail of part of a VS of lamina of a leaf of *Prunus domestica* showing part of the vein network in LS and VS (x 180)

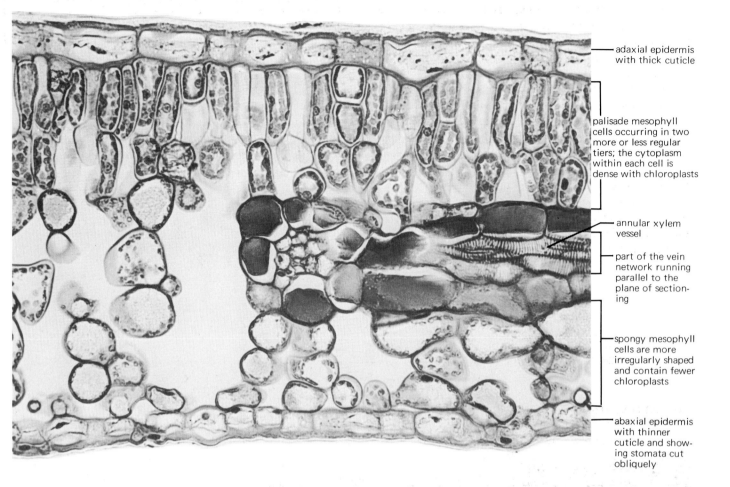

adaxial epidermis with thick cuticle

palisade mesophyll cells occurring in two more or less regular tiers; the cytoplasm within each cell is dense with chloroplasts

annular xylem vessel

part of the vein network running parallel to the plane of sectioning

spongy mesophyll cells are more irregularly shaped and contain fewer chloroplasts

abaxial epidermis with thinner cuticle and showing stomata cut obliquely

Figure 2.4(ii) Drawing of part of the lamina section above

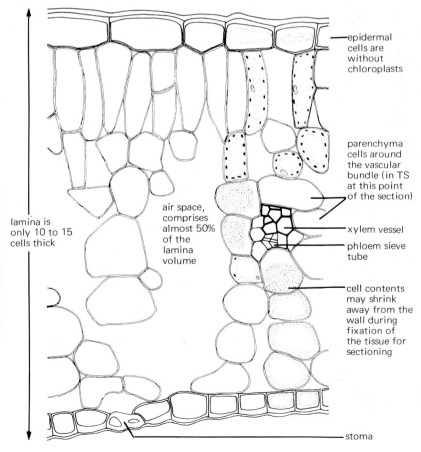

epidermal cells are without chloroplasts

parenchyma cells around the vascular bundle (in TS at this point of the section)

xylem vessel

phloem sieve tube

cell contents may shrink away from the wall during fixation of the tissue for sectioning

lamina is only 10 to 15 cells thick

air space, comprises almost 50% of the lamina volume

stoma

Secretion in plants is a common phenomenon. In certain cases particular secretory spaces or canals develop (e.g. resin canals of *Pinus*, see page 35). Mostly secretion occurs either in the cytoplasm or from the cytoplasm to the vacuole, or alternatively from the cytoplasm into the cell wall.

E.g. *Tannins* are complex aromatic compounds that have a bitter taste and form dark-coloured complexes within cells. They occur in bark, leaves, and unripe fruits. Those in plums are decomposed as the fruits ripen. Tannins precipitate proteins in tanning leather. Bark of oaks and chestnuts have been used as commercial sources. Tannins are more abundant in cell walls than in the cytoplasm.

E.g. *Anthocyanins* are water-soluble pigments that occur in vacuoles. They are responsible for the red, pink, violet, and blue colours of flowers. They may also occur elsewhere in plants, e.g. beetroot tissue, and in leaves deficient in nitrogen. They are pH indicators and vary in colour with the acidity or alkalinity of the vacuole solution.

31

Stomata

Stomata consist of two almost sausage-shaped *guard cells* in the epidermis. Guard cells separate to form a *pore* between them. They occur in leaves, also in stems and in the parts of flowers. Stomata occur mostly in the lower epidermis of dorsiventral leaves, but are often evenly distributed in the leaves of monocotyledons (e.g. *Allium cepa* (Onion), *Zea mays* (Maize)). The stomata of most species are open in daylight and are closed in the dark. When the guard cells are turgid the stoma is open; when the guard cells are flaccid the pore is closed. Because of the unique shape of the guard cells (the walls are thin laterally but thickened on the outer and inner faces) and of the direction in which the strands of cellulose fibres are laid down in the walls, the guard cells bulge laterally into the *subsidiary cells* as they become turgid, and the pore appears.

Stomata will close in the light if the leaf begins to wilt or if a high concentration of carbon dioxide occurs in the air space of the leaf. Many hypotheses have been suggested relating changes in light, temperature, carbon dioxide concentration, and water supply to the changes in osmotic potential of the guard cells that cause change in size of the stomatal pore. No single hypothesis of the opening and closing mechanism has proved universally adequate. Guard cells are the only cells of the epidermis to contain chloroplasts and therefore carry out photosynthesis in the light.

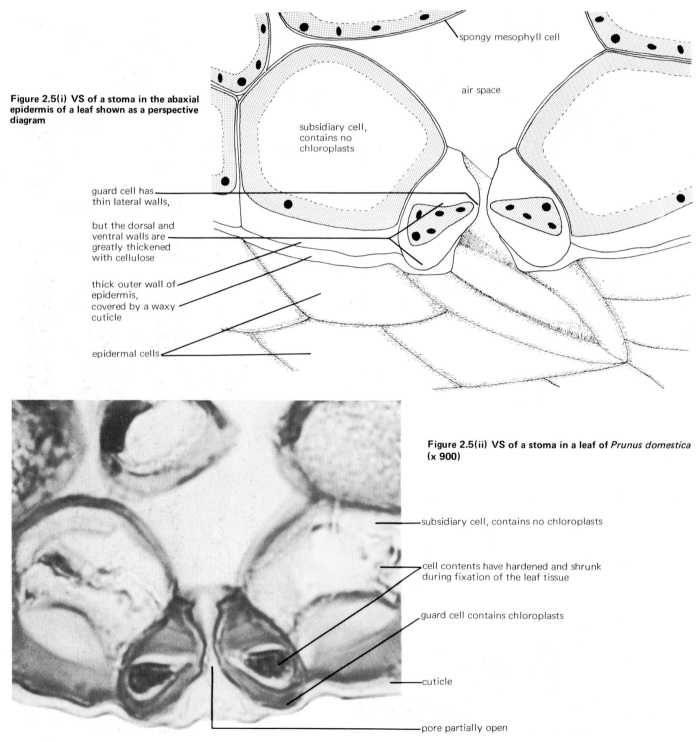

Figure 2.5(i) VS of a stoma in the abaxial epidermis of a leaf shown as a perspective diagram

spongy mesophyll cell

air space

subsidiary cell, contains no chloroplasts

guard cell has thin lateral walls,

but the dorsal and ventral walls are greatly thickened with cellulose

thick outer wall of epidermis, covered by a waxy cuticle

epidermal cells

Figure 2.5(ii) VS of a stoma in a leaf of *Prunus domestica* (x 900)

subsidiary cell, contains no chloroplasts

cell contents have hardened and shrunk during fixation of the leaf tissue

guard cell contains chloroplasts

cuticle

pore partially open

Figure 2.6(i) Surface view of the abaxial epidermis of a dicotyledonous leaf, *Ligustrum vulgare* **(Privet) (x 300)**

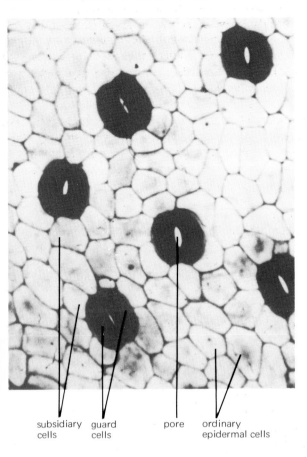

subsidiary · guard · pore · ordinary
cells · cells · · epidermal cells

The elliptical stomata of the dicotyledonous leaf are scattered evenly in the epidermis. Sections through leaves frequently cut through stomata, but only rarely do such sections occur through the centre of the pore at right angles to the long axis, providing the easily interpreted view shown in Figure 2.5(ii) opposite. Mostly stomata appear as minor irregularities in the continuity of the epidermis when seen in section, and careful observation of more than one section may be necessary to find views similar to those illustrated here. Stomata of many monocotyledons with bayonet-shaped leaves occur in regular pattern, in rows parallel to the length of the leaf (e.g. Gramineae (Grasses) and Liliaceae (Lily) families).

Figure 2.6(ii) Surface view of a closed stoma of *Buxus sempervirens* **(Box), stained to show the nucleus in each cell (x 550)**

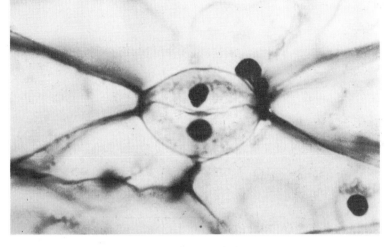

Diffusion of gases in and out of the leaf occurs through the stomata. The total pore area of open stomata in a leaf is between 0.5% and 2.0% of the total area of the leaf, although the frequency of the stomata varies between 10 and 400 per square millimetre.

Despite this small area of pore, carbon dioxide can diffuse into the leaf in sunlight almost as rapidly as if the whole epidermal barrier were not there. At the same time water vapour is lost from the leaf by diffusion. But stomata do prevent severe desiccation of the mesophyll cells when the stomata are closed in the wilting leaf.

Figure 2.6(iii) Surface view of the epidermis of a monocotyledonous leaf, *Iris* **sp. (Iris) (x 250)**

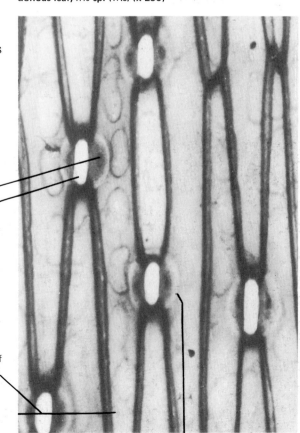

nucleus of the guard cell
pore wide open

Figure 2.6(iv) Drawing of a VS through a stoma in the epidermis of *Iris* **sp. leaf**

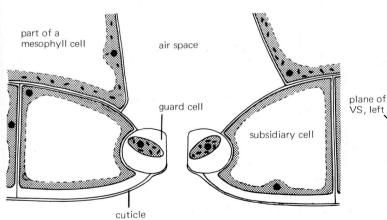

part of a mesophyll cell · air space

guard cell

subsidiary cell

plane of VS, left

cuticle

the lateral walls of the guard cell of *Iris* appear as a faint line in surface view because adjoining cells arch over the guard cells; when the pore is fully open the guard cells are deeply embedded in the subsidiary cells

33

Figure 2.7(i) Leaf of *Ilex aquifolium* (Holly) (life size)

VS of petiole seen in Figure 2.7(iii)

VS seen in Figure 2.7(ii)

Evergreen leaves

The Holly *(Ilex aquifolium)* is an evergreen tree. It produces new leaves in the spring. These survive through one or two winters and fall off in spring time. In winter time water uptake by roots is often curtailed, particularly when the ground is frozen. *Ilex aquifolium* leaves would face the dangers of desiccation if water loss by transpiration were not reduced. The pronounced cuticle of the upper and lower epidermis, together with the possession of an outer antechamber to the stomatal pore, are advantageous in reducing water vapour loss, and are typical of *xerophytic plants* (plants adapted to life in dry places).

Figure 2.7(ii) VS of part of the lamina of a leaf of *Ilex aquifolium* (x 20)

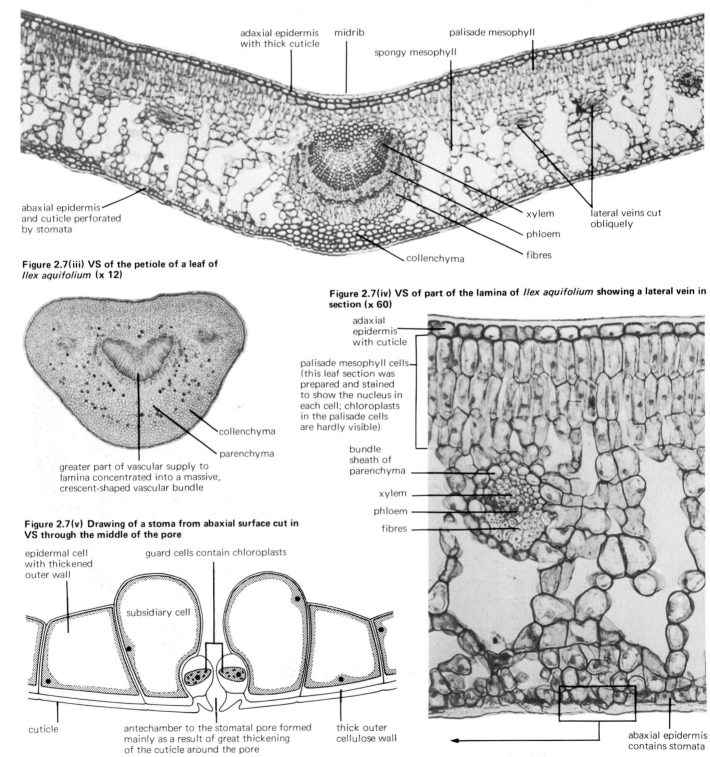

adaxial epidermis with thick cuticle

midrib

palisade mesophyll

spongy mesophyll

abaxial epidermis and cuticle perforated by stomata

xylem

lateral veins cut obliquely

phloem

collenchyma

fibres

Figure 2.7(iii) VS of the petiole of a leaf of *Ilex aquifolium* (x 12)

collenchyma

parenchyma

greater part of vascular supply to lamina concentrated into a massive, crescent-shaped vascular bundle

Figure 2.7(iv) VS of part of the lamina of *Ilex aquifolium* showing a lateral vein in section (x 60)

adaxial epidermis with cuticle

palisade mesophyll cells (this leaf section was prepared and stained to show the nucleus in each cell; chloroplasts in the palisade cells are hardly visible)

bundle sheath of parenchyma

xylem

phloem

fibres

abaxial epidermis contains stomata

Figure 2.7(v) Drawing of a stoma from abaxial surface cut in VS through the middle of the pore

epidermal cell with thickened outer wall

guard cells contain chloroplasts

subsidiary cell

cuticle

antechamber to the stomatal pore formed mainly as a result of great thickening of the cuticle around the pore

thick outer cellulose wall

34

Figure 2.8(i) A short length of a shoot of *Pinus sylvestris* (Scots Pine) showing attachment of needles

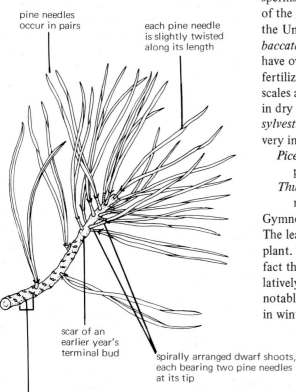

pine needles occur in pairs

each pine needle is slightly twisted along its length

scar of an earlier year's terminal bud

spirally arranged dwarf shoots, each bearing two pine needles at its tip

scars of dwarf shoots that have been shed with pine needles attached

The Scots Pine *(Pinus sylvestris)* bears needles as leaves. This plant is classified as a *gymnosperm* or naked-seeded plant (seed-bearing plants include the angiosperms (flowering plants) and gymnosperms.) Most of the living representatives of the gymnosperms are conifers (softwood trees) of which three are native to the United Kingdom *(Pinus sylvestris, Juniperus communis* (Juniper), and *Taxus baccata* (Yew)), but many other species have been introduced. All gymnosperms have ovules exposed on the scales of immature female cones. These ovules are fertilized by pollen carried by wind from male cones. Except at pollination the scales are tightly shut as the seeds develop, but when ripe the scales move apart in dry weather and the winged seeds are blown away. (See life-cycle of *Pinus sylvestris,* page 64). Softwood trees grow fast on poor soils and have become very important in world economy. For example:

Picea abies (Norway Spruce) for whitewood furniture, wood pulp for paper production, and as Christmas trees.

Thuja plicata (Western Red Cedar) in timber buildings because the timber is rot-resistant.

Gymnosperms bear leaves with a relatively small surface area to volume ratio. The leaves are very numerous and much transpiration occurs from the whole plant. The water relations of the whole plants are further complicated by the fact that the water-conducting tissue is of tracheids only (see page 7) and is relatively resistant to upward water flow, and the conifers (with few exceptions, notably Larch) are evergreen trees so that the leaves endure critical conditions in winter. The leaves show many features typical of xerophytic plants.

Figure 2.8(ii) TS of a needle of *Pinus sylvestris* (x 120)

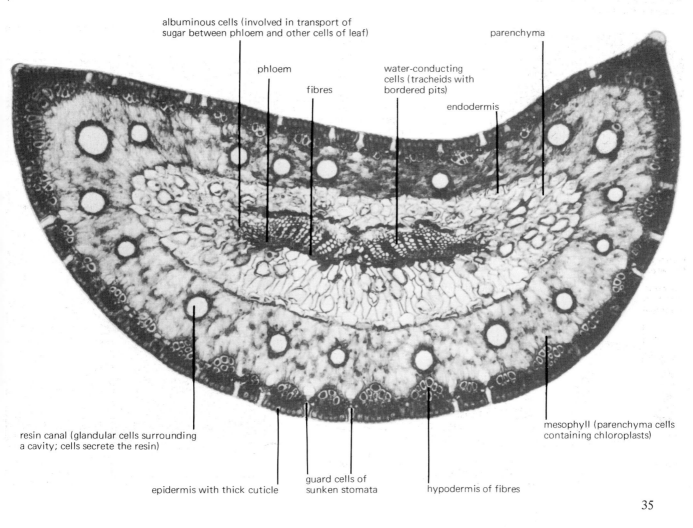

albuminous cells (involved in transport of sugar between phloem and other cells of leaf)

parenchyma

phloem

fibres

water-conducting cells (tracheids with bordered pits)

endodermis

resin canal (glandular cells surrounding a cavity; cells secrete the resin)

mesophyll (parenchyma cells containing chloroplasts)

epidermis with thick cuticle

guard cells of sunken stomata

hypodermis of fibres

35

Leaves of monocotyledons

The leaves of monocotyledons vary in form and structure and some resemble the leaves of dicotyledons in being differentiated into leaf blade and petiole. The majority of species of monocotyledons have leaves differentiated into *blade* and *sheath*. The sheaths partially overlap younger leaves and the growing point of the stem. The leaf blade is usually narrow and the venation is typically *parallel*. Those leaf blades that are borne vertically are not differentiated into palisade and spongy mesophyll within, and are called *unifacial* or *isobilateral* leaves. *Iris* spp. has a unifacial leaf.

Figure 2.9(i) *Iris* **sp. (Iris) showing rhizome and aerial system**

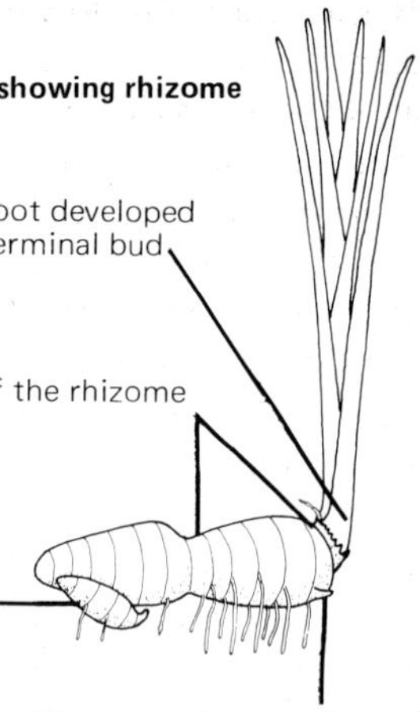

aerial shoot developed from a terminal bud

one year's growth of the rhizome

lateral branch of the rhizome

lateral bud containing horizontal growth of the rhizome

Figure 2.9(ii) A leaf of *Iris* sp. (Yellow Flag)

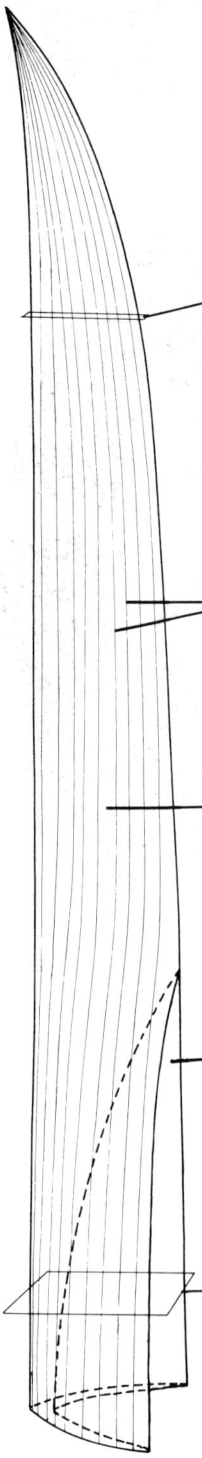

Figure 2.9(iii) TS of part of a leaf blade of *Iris* sp. (x 35)

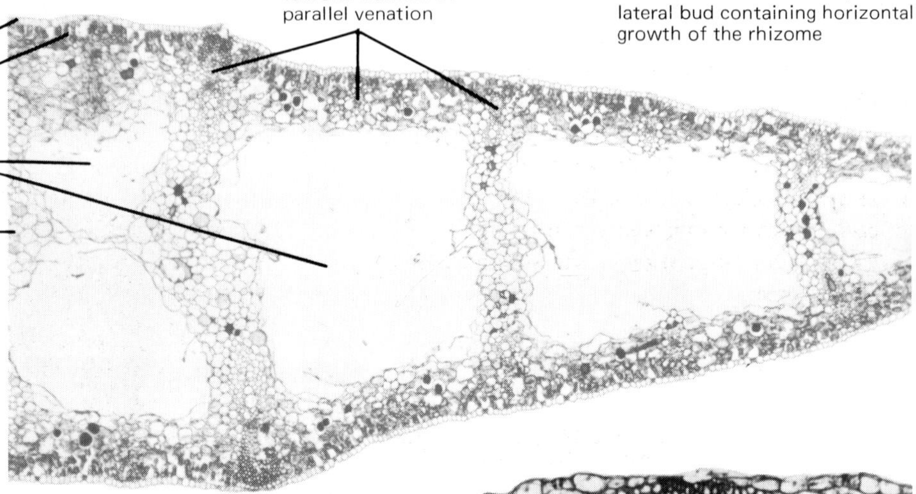

vascular bundles of parallel venation

epidermis

mesophyll

air spaces

parenchyma

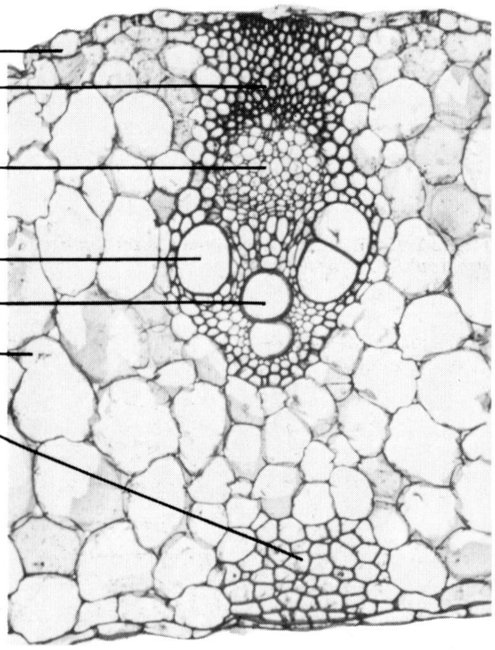

main veins run parallel in the leaf

leaf blade

leaf sheath

epidermis

fibres

phloem

metaxylem vessel

protoxylem

mesophyll

fibres

Figure 2.9(iv) TS of part of a leaf blade of *Zea mays* (Maize) (x 130)

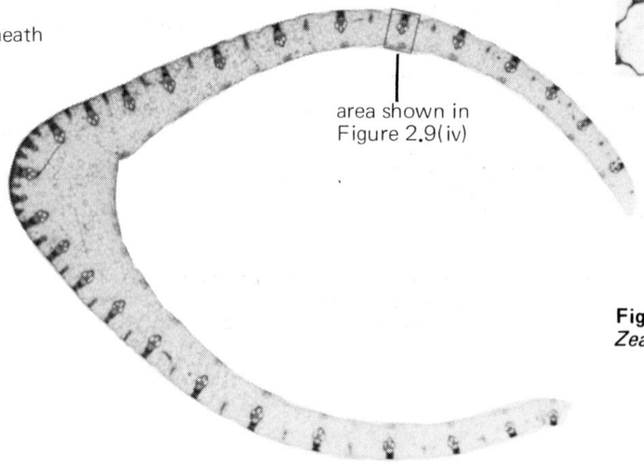

area shown in Figure 2.9(iv)

Figure 2.9(v) TS of the leaf sheath of *Zea mays* (x 10)

Marram grass *(Ammophila arenaria)* is the dominant species of plant colonizing sand dunes around the coasts of western Europe. It endures extremely dry conditions. Small heaps of sand collect around plants of *Ammophila* and fresh shoots grow up through the sand that collects over them. The larger tufts of leaves accelerate deposition of sand from the wind and shifting dunes gradually gather a dense clothing of vegetation.

Ammophila arenaria is a *xerophyte* since it grows in an arid habitat and can decrease transpiration to a minimum under conditions of water deficiency. The leaf has a small ratio of external leaf surface to its volume when rolled up circular in section. The abaxial epidermis has a thick cuticle and no stomata, these being confined to the furrows of the inner surface where additional protection is provided by the stiff, interlocking hairs. The exceptionally large epidermal cells (hinge cells) at the base of each furrow are thin-walled and shrink quickly when transpiration is excessive, making the leaf tubular.

Figure 2.10(i) *Ammophila arenaria* (Marram Grass) growing on sand dunes

Figure 2.10(ii) TS of the leaf of *Ammophila arenaria* (x 50)

Figure 2.10(iii) Tissue map of part of the lamina of *Ammophila arenaria*

abaxial epidermis with thick cuticle

adaxial epidermis

fibres

position of hinge cells

unicellular hair, outgrowth of an epidermal cell

sheath of small fibres
xylem   } vascular bundle
phloem

mesophyll cells, thin-walled and containing chloroplasts (stomata occur in the epidermis of regions of mesophyll cells)

Figure 2.11(i) TS of a vegetative bud of *Acer pseudoplatanus* **(Sycamore) (x 20)**

Buds

Woody plants form resting buds at some stage in the annual cycle of growth. In temperate zone species, buds are formed in late summer and the whole tree or shrub enters a dormant phase. Resting buds are also formed by tropical species, but not all the shoots of the tree may enter the dormant phase simultaneously. Resting buds consist of *leaf primordia* (partly developed foliage leaves) around the stem apex and are surrounded by *bud scales.* In most woody species the shortening of days in late summer is the stimulus that triggers off the formation of resting buds and the onset of dormancy. In *Acer pseudoplatanus* (Sycamore) this photoperiodic response is determined by the day length to which the mature leaves are exposed, rather than the shoot apical region itself being sensitive.

Once they are formed the dormant buds require winter-chilling before growth can be resumed. Most woody species need from two to ten weeks at about 0 to 5 °C to complete the bud dormancy stage. Then the time of bud-break is determined by rising temperature in the spring.

folded, incompletely developed foliage leaf

glandular hairs

vascular bundles

Figure 2.11(ii) Foliage leaf of *Acer pseudoplatanus.*
A copy of this leaf may be traced, cut out and folded as shown in Figure 2.11(iii) to aid your understanding of the arrangement of immature leaves in the bud

scale leaf (modified leaf base), with dark, hard outer surface, protects against frost damage, mechanical injury or attack by predators

Figure 2.11(iii) Lateral bud of *Acer pseudoplatanus* **cut open to show one of the folded, immature foliage leaves** *in situ*

Sycamore leaves are often stained with black patches caused by Tar-spot Fungus which causes little damage to the trees. The fungus is poisoned by sulphur dioxide gas in polluted air, and is not seen on trees in urban and industrial areas.

5–7 veins of almost equal prominence

At bud-break the bud scales fall off, leaving a ring scar at the base of the new growth. The internodes expand and the new leaves unfold and grow. As these leaves appear from the bud they are often covered in fine *epidermal hairs.* These trap moist air above the surface of the lamina at a time when the cuticle is not fully developed. The Sycamore leaf has a row of short *glandular hairs* on the adaxial surface along the line of each main vein.

Leaf fall

The fall of leaves is preceded by the different-
iation of a definite separating layer *(abscission
layer)* in the cells at the base of the leaf stalk.
The process involves the formation of several
layers of small cells by cell division, and then
the gelatinization of the middle lamellae of
these cells so that the layer disintegrates, resulting in abscission.

Beneath the abscission layer on the stem side a
protective layer is formed by lignification and
suberization of the underlying cells. After leaf
fall the scar becomes covered by a layer of cork
which becomes continuous with that of the
stem. The cork layer does not seal the xylem
vessels. These become plugged by tyloses (in-
growths from surrounding living cells, see page
21).

The process of abscission layer formation is in-
hibited by the hormone auxin *(indole acetic
acid)*. Young leaves are the site of active auxin
formation. As leaves age the export of auxin
from them progressively decreases. There is
some evidence that other endogenous hormones
(including ethylene and abscisic acid) may also
be involved in the process of leaf fall.

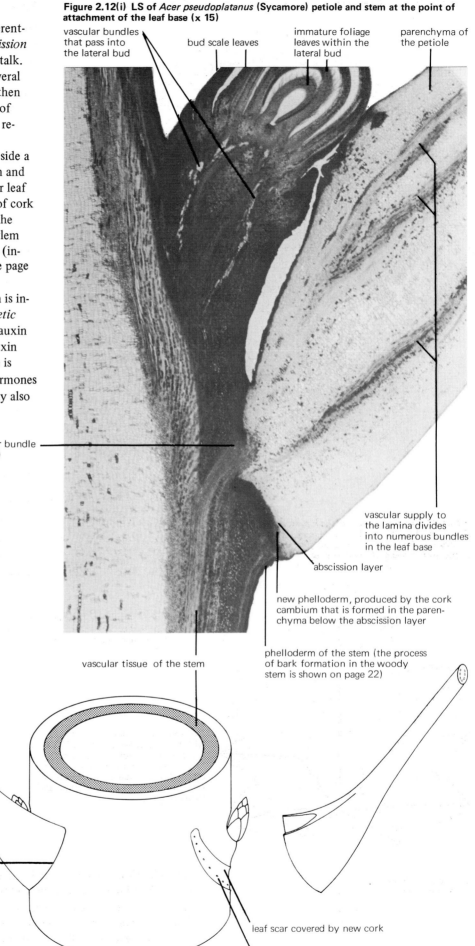

Figure 2.12(i) LS of *Acer pseudoplatanus* (Sycamore) petiole and stem at the point of attachment of the leaf base (x 15)

vascular bundles that pass into the lateral bud

bud scale leaves

immature foliage leaves within the lateral bud

parenchyma of the petiole

leaf trace (vascular bundle supplying the leaf)

vascular supply to the lamina divides into numerous bundles in the leaf base

abscission layer

new phelloderm, produced by the cork cambium that is formed in the paren-chyma below the abscission layer

phelloderm of the stem (the process of bark formation in the woody stem is shown on page 22)

vascular tissue of the stem

Figure 2.12(ii) Diagram showing leaf fall

Around the lateral bud the petiole forms a secure attach-ment to the stem. The formation of the abscission layer is initiated by the lower temperature and shorter days of autumn. This abscission layer forms first in the outer tissue and moves inwards to converge on the central vascular tissue.

leaf scar covered by new cork

broken vascular bundle

Section 3: The root-anchorage and absorption

Introduction

The **root** originates from the *radicle* of the seedling at germination. If this main root persists and grows down, a *tap-root system* is formed.

In the grasses and in other monocotyledonous families of flowering plants the main root rapidly withers, and numerous *adventitious roots* develop from the base of the stem to form a *fibrous root system*.

Figure 3.1(i) Early stages in the germination of *Zea mays* (Maize) fruit

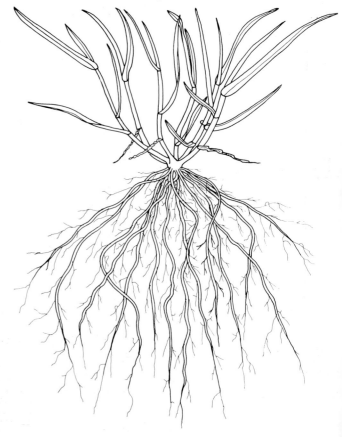

1 emergence of the coleoptile and the radicle

2 adventitious roots appear at the base of the coleoptile

3 the coleoptile splits open at the tip and the first leaf emerges in the light

the tap-root withers and the fibrous root system develops

Figure 3.1(ii) Tap-root system of *Capsella bursa-pastoris* (Shepherd's Purse) (life size)

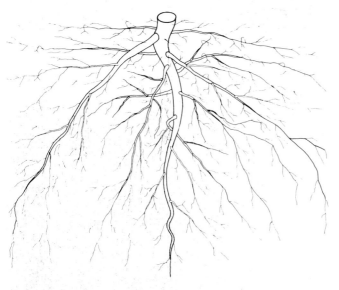

In most dicotyledons and gymnosperms the root system consists of a tap-root from which lateral roots arise. Roots bear no appendages comparable to leaves or buds which have a superficial origin on the stem. Lateral roots originate deep in the tissue near the centre of the root (see Figure 3.7(i), page 46). Roots also possess a root cap at the tip (Figure 3.2(iii) opposite) not found on stems. Roots do not have stomata.

The root system grows in the soil and provides *anchorage* for the aerial system. Total root length of individual plants must be measured in kilometres: for example, a wheat plant may have 70 kilometres of roots. Roots of trees and shrubs grow less rapidly than those of herbaceous plants, but a mature tree possesses many kilometres of roots spread over an area of soil as great as, or greater than, the area covered by the branches. The older portions of the roots, nearest the surface, usually undergo *secondary thickening*. This increases their strength and therefore the support they provide to the aerial system. Roots are also a site of *starch storage*.

The *uptake of water and of salts as ions* is carried out by the new extremities of the root system whilst still in the process of primary growth. The continuing growth of root tips results in new areas of soil being contacted by root tissue. *Root hairs* increase the surface area of roots in this region. Root growth and root metabolism (this includes the important process of synthesis of amino acids for the whole plant) require air, and roots in trodden-down or water-logged soil quickly die. Soil air is as vital for root growth as is soil water.

Figure 3.1(iii) A complete plant of *Poa annua* (Annual Meadow Grass) showing the large fibrous root system (life size)

40

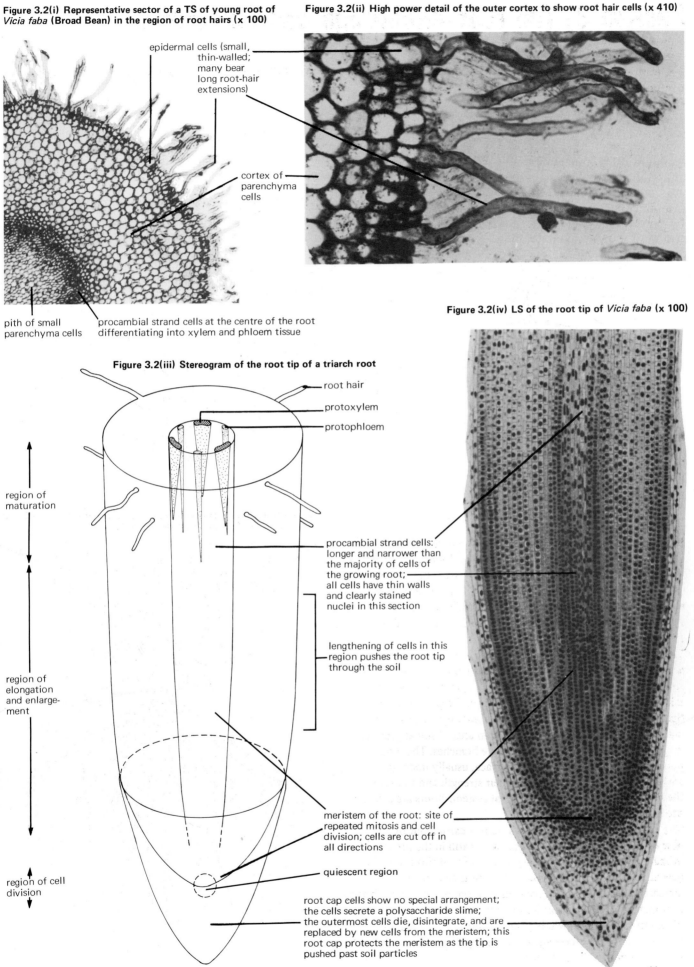

Figure 3.2(i) Representative sector of a TS of young root of *Vicia faba* **(Broad Bean) in the region of root hairs (x 100)**

Figure 3.2(ii) High power detail of the outer cortex to show root hair cells (x 410)

epidermal cells (small, thin-walled; many bear long root-hair extensions)

cortex of parenchyma cells

pith of small parenchyma cells

procambial strand cells at the centre of the root differentiating into xylem and phloem tissue

Figure 3.2(iv) LS of the root tip of *Vicia faba* **(x 100)**

Figure 3.2(iii) Stereogram of the root tip of a triarch root

root hair

protoxylem

protophloem

region of maturation

region of elongation and enlarge-ment

region of cell division

procambial strand cells: longer and narrower than the majority of cells of the growing root; all cells have thin walls and clearly stained nuclei in this section

lengthening of cells in this region pushes the root tip through the soil

meristem of the root: site of repeated mitosis and cell division; cells are cut off in all directions

quiescent region

root cap cells show no special arrangement; the cells secrete a polysaccharide slime; the outermost cells die, disintegrate, and are replaced by new cells from the meristem; this root cap protects the meristem as the tip is pushed past soil particles

41

Figure 3.3.(i) Drawing of part of a root to show the region of mature primary growth and region of root hairs in relation to the root cap

developing exodermis

central vascular tissue

withered ⎤
mature ⎬ root hairs
developing ⎦

region of elongation and expansion of cells

root cap

Figure 3.3.(ii) TS of mature root of *Ranunculus* sp. (Buttercup) (x 100)

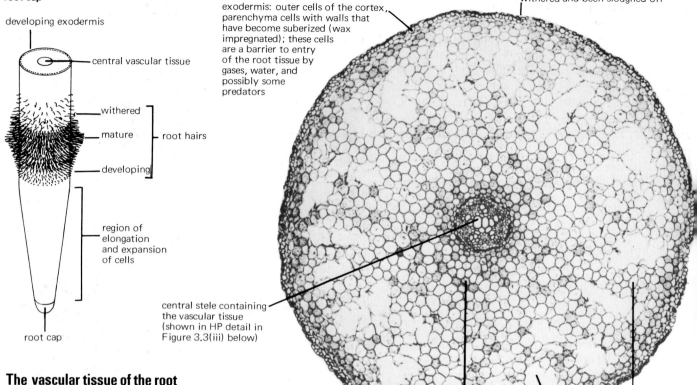

epidermis from which root hairs have withered and been sloughed off

exodermis: outer cells of the cortex, parenchyma cells with walls that have become suberized (wax impregnated); these cells are a barrier to entry of the root tissue by gases, water, and possibly some predators

central stele containing the vascular tissue (shown in HP detail in Figure 3.3(iii) below)

starch grains in evidence, particularly in the cells surrounding the stele

cortex of parenchyma cells (large air spaces occur, and facilitate diffusion of gases in the root)

The vascular tissue of the root

Comparing the position of xylem and phloem in the root and the stem reveals two important differences. In the root:

(a) Vascular tissue occurs centrally, in a compact *stele*. Soil is an entirely supporting medium through which the flexible root grows. (Air is a non-supporting medium in which the rigid stem supports itself, having girder-like vascular bundles just below the epidermis).

(b) Protoxylem and protophloem alternate around the stele, rather than in collateral bundles as in the stem. The root shown in section here has five protoxylem groups and is therefore termed *pentarch* (cf. roots shown in Figures 3.4(i) and (iii) opposite).

Metaxylem is formed on the inner side of the protoxylem (centripetally, called *exarch* arrangement of xylem) in contrast to the metaxylem of the stem (formed centrifugally, called *endarch* arrangement; see page 45).

Figure 3.3.(iii) High power detail of the stele of *Ranunculus* sp. root (x 400)

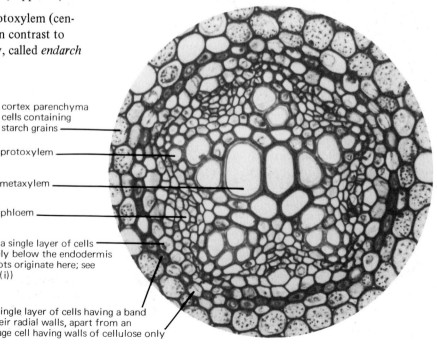

Figure 3.3.(iv) Drawing of a representative portion of the stele

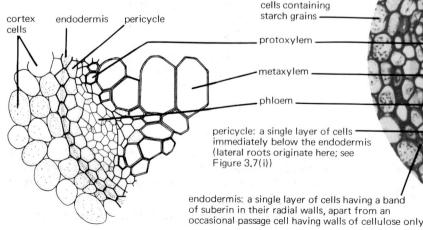

cortex cells   endodermis   pericycle

cortex parenchyma cells containing starch grains

protoxylem

metaxylem

phloem

pericycle: a single layer of cells immediately below the endodermis (lateral roots originate here; see Figure 3.7(i))

endodermis: a single layer of cells having a band of suberin in their radial walls, apart from an occasional passage cell having walls of cellulose only

The root in dicotyledons

In dicotyledons the number of protoxylem groups that develop remains small. The pentarch arrangement (Figure 3.3(iii) opposite) is the highest number found in dicotyledons. Roots that have only two protoxylem groups are called *diarch*, and are typically found in *Nicotinia* spp. (Tobacco), *Beta* spp. (Beet), and *Daucus* spp. (Carrot). *Triarch* roots typically occur in *Pisum* spp. (Pea). *Tetrarch* roots occur typically in *Vicia* spp. (Vetch), and in *Ranunculus* spp. (Buttercup).

Variation in number of protoxylem groups often occurs in roots of the same plant. The stele in the radicle usually shows the number of protoxylem groups typical of the plant. The later formed roots may vary and show a number of protoxylem groups related to the size of the particular root apex.

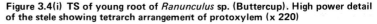

Figure 3.4(i) TS of young root of *Ranunculus* sp. (Buttercup). High power detail of the stele showing tetrarch arrangement of protoxylem (x 220)

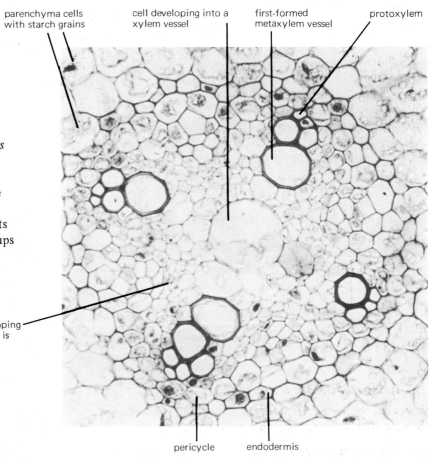

parenchyma cells with starch grains

cell developing into a xylem vessel

first-formed metaxylem vessel

protoxylem

protophloem (developing phloem — formation is centripetal)

pericycle

endodermis

Figure 3.4(ii) TS of young root of *Helianthus* sp. (Sunflower). High power detail of the stele showing the diarch arrangement of protoxylem (x 220)

protoxylem vessels

metaxylem vessels

cell developing into a xylem vessel

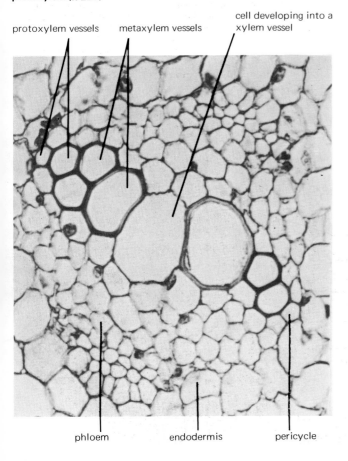

phloem

endodermis

pericycle

As the primary vascular tissue is laid down in the young root a *casparian strip* appears on the radial and cross walls of the *endodermal cells*. This is a narrow band of cellulose impregnated with suberin and lignin. This impervious material is a barrier to the free passage of water and of salts in solution through all the fine free spaces in the cellulose walls of cells.

Figure 3.4(iii) An endodermal cell

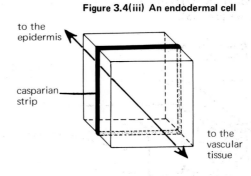

to the epidermis

casparian strip

to the vascular tissue

As a consequence all material entering the root must pass through the cell protoplast at the endodermis. In older roots the thickening in the endodermis walls may increase greatly and occur also along the inner tangential wall. This forms a complete barrier to entry into the stele, except at the position of the *passage cells*, which occur opposite the protoxylem groups.

43

The root in monocotyledons

In the roots of monocotyledons the number of protoxylem groups is always large. The *Iris* sp. root shown here has twelve protoxylem groups, and is described as *polyarch*. Also, the endodermis cells surrounding the stele of monotyledons typically have greatly thickened and suberized radial and inner walls. The occasional passage cell is conspicuous by its dense living contents and thin walls.

Figure 3.5(i) Part of a mature root of *Iris* sp. (x 35)

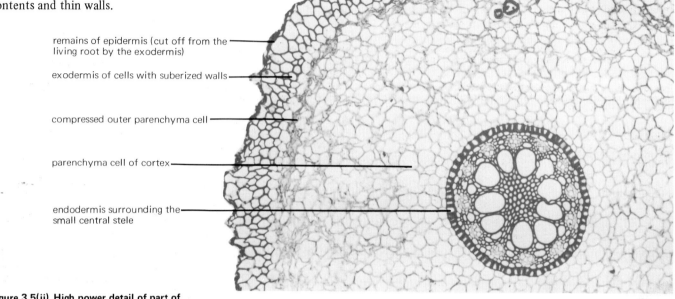

remains of epidermis (cut off from the living root by the exodermis)

exodermis of cells with suberized walls

compressed outer parenchyma cell

parenchyma cell of cortex

endodermis surrounding the small central stele

Figure 3.5(ii) High power detail of part of the stele of *Iris* sp. root (x 400)

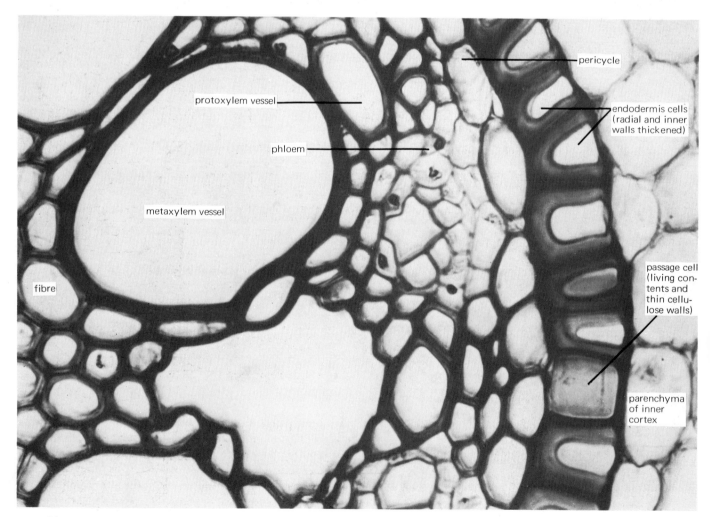

pericycle

protoxylem vessel

phloem

endodermis cells (radial and inner walls thickened)

metaxylem vessel

fibre

passage cell (living contents and thin cellulose walls)

parenchyma of inner cortex

Transition–root becomes stem

The *transition zone* occurs between the root and the stem at soil level. It is a narrow region, and the organization of it differs in different species of plants.

In the root xylem and phloem occur separately, on different radii. Xylem development is exarch so that protoxylem elements are on the outside. The vascular tissue occurs in the centre of the root in a compact stele with no pith.

In the stem xylem and phloem occur in collateral bundles. Xylem development is endarch so that protoxylem groups are at the inner-most edge of each vascular bundle. The vascular bundles occur around the margin of the stem with a large central pith, particularly in dicotyledonous stems. The structure of the stem is usually made more complicated by the emergence of numerous leaf trace bundles from the ring of bundles.

The transition zone is a *complex region of interchange.*

Figure 3.6(i) 'Exploded' stereogram to show the transition zone of *Helianthus annuus* (Sunflower) (follow the sequence of transition from root to stem)

Figure 3.6(ii) TS of (A) vascular bundles of stem, (B) transition zone, and (C) stele of the root, in *Helianthus annuus* (x 50)

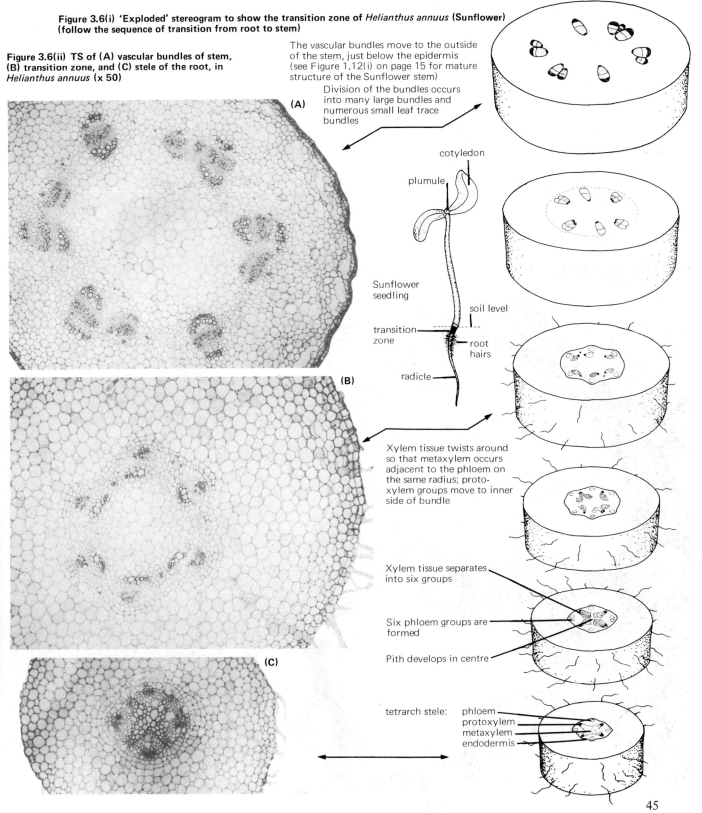

(A)

The vascular bundles move to the outside of the stem, just below the epidermis (see Figure 1.12(i) on page 15 for mature structure of the Sunflower stem)

Division of the bundles occurs into many large bundles and numerous small leaf trace bundles

cotyledon

plumule

Sunflower seedling

soil level

transition zone

root hairs

radicle

(B)

Xylem tissue twists around so that metaxylem occurs adjacent to the phloem on the same radius; proto-xylem groups move to inner side of bundle

Xylem tissue separates into six groups

Six phloem groups are formed

Pith develops in centre

(C)

tetrarch stele: phloem
protoxylem
metaxylem
endodermis

Lateral roots

In complete contrast with shoot structures such as leaves and lateral stems, lateral roots arise at a considerable distance from the apex. Furthermore, the lateral root meristem develops in the *pericycle* layer within the stele at the centre of the root, rather than having a superficial origin.

Lateral roots are initiated by periclinal (parallel with the circumference) and anticlinal (perpendicular to the surface) divisions in a group of pericycle cells opposite a protoxylem group. These new cells add to the volume of the root causing a lump at the surface of the root very early in development. The new lateral root penetrates outwards through the cortex displacing parenchyma cells, and finally bursts out through the exodermis and epidermis into the soil.

Xylem and phloem differentiate in the lateral root and become connected with the equivalent elements in the parent root. The lateral root develops in the same way as the main root from which it was derived. For example the lateral root bears root hairs behind the region of elongation at the root tip, and the lateral root ultimately develops further lateral roots. A consequence of root development is the continuous creation of new absorbing surfaces in contact with fresh soil.

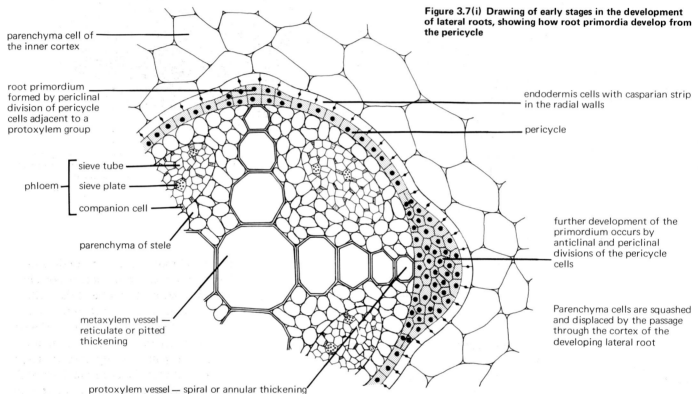

Figure 3.7(i) Drawing of early stages in the development of lateral roots, showing how root primordia develop from the pericycle

parenchyma cell of the inner cortex

root primordium formed by periclinal division of pericycle cells adjacent to a protoxylem group

phloem — sieve tube / sieve plate / companion cell

parenchyma of stele

metaxylem vessel — reticulate or pitted thickening

protoxylem vessel — spiral or annular thickening

endodermis cells with casparian strip in the radial walls

pericycle

further development of the primordium occurs by anticlinal and periclinal divisions of the pericycle cells

Parenchyma cells are squashed and displaced by the passage through the cortex of the developing lateral root

Figure 3.7(ii) TS of root of *Vicia faba* (Broad Bean) showing an early stage in the formation of a lateral root (x 80)

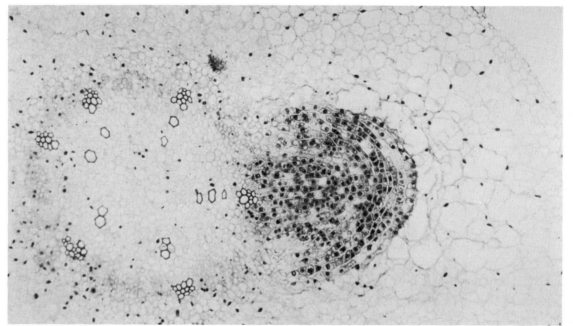

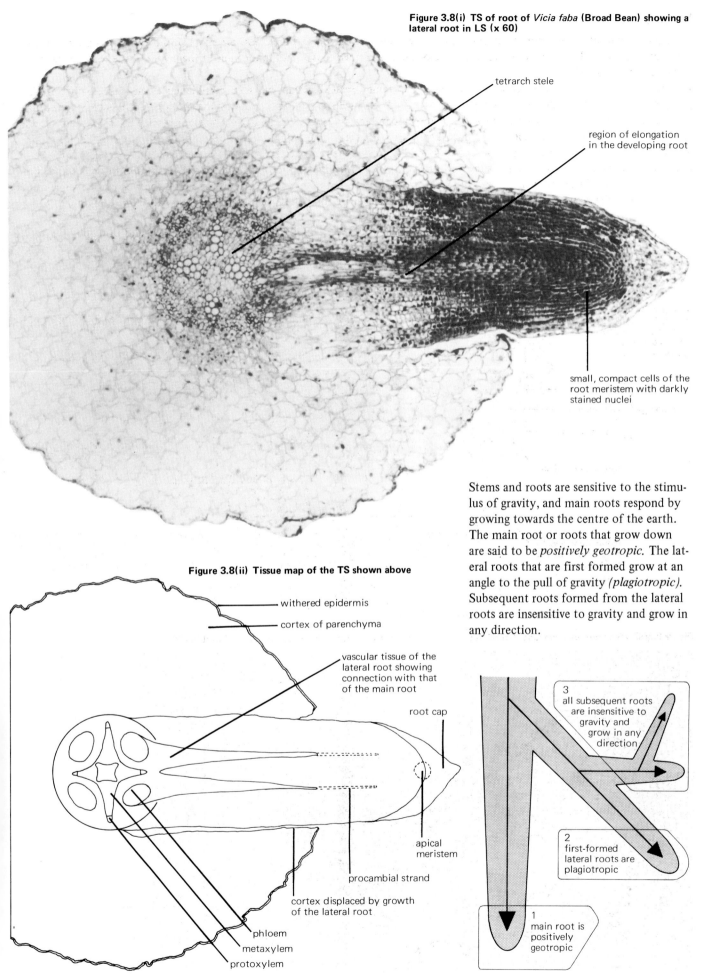

Figure 3.8(i) TS of root of *Vicia faba* (Broad Bean) showing a lateral root in LS (x 60)

tetrarch stele

region of elongation in the developing root

small, compact cells of the root meristem with darkly stained nuclei

Figure 3.8(ii) Tissue map of the TS shown above

withered epidermis

cortex of parenchyma

vascular tissue of the lateral root showing connection with that of the main root

root cap

apical meristem

procambial strand

cortex displaced by growth of the lateral root

phloem

metaxylem

protoxylem

Stems and roots are sensitive to the stimulus of gravity, and main roots respond by growing towards the centre of the earth. The main root or roots that grow down are said to be *positively geotropic*. The lateral roots that are first formed grow at an angle to the pull of gravity *(plagiotropic)*. Subsequent roots formed from the lateral roots are insensitive to gravity and grow in any direction.

3
all subsequent roots are insensitive to gravity and grow in any direction

2
first-formed lateral roots are plagiotropic

1
main root is positively geotropic

47

The older root

Secondary growth in roots consists of formation of secondary xylem and phloem and of protective cork *(periderm),* and occurs in roots of dicotyledons, although in some of them the roots show primary growth only throughout life. Monocotyledons lack secondary growth altogether.

Vascular cambium appears between the primary phloem and xylem in the stele. Then parenchyma cells and pericycle cells around the protoxylem groups become meristematic and connect up with the vascular cambium below the phloem. Thus a corrugated cylinder of vascular cambium around the xylem is formed within the stele. This cambium cylinder has the same shape in cross-section as the outline of the primary xylem.

Figure 3.9(i) TS of the stele of a root of *Vicia faba* (Broad Bean) showing the development of vascular cambium on the inner face of the phloem (x 140)

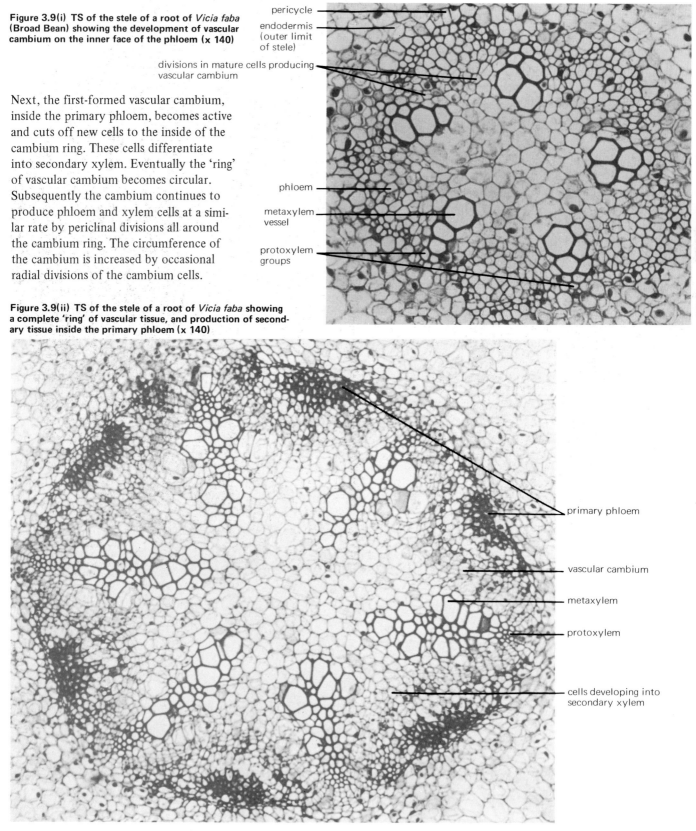

pericycle

endodermis (outer limit of stele)

divisions in mature cells producing vascular cambium

phloem

metaxylem vessel

protoxylem groups

Next, the first-formed vascular cambium, inside the primary phloem, becomes active and cuts off new cells to the inside of the cambium ring. These cells differentiate into secondary xylem. Eventually the 'ring' of vascular cambium becomes circular. Subsequently the cambium continues to produce phloem and xylem cells at a similar rate by periclinal divisions all around the cambium ring. The circumference of the cambium is increased by occasional radial divisions of the cambium cells.

Figure 3.9(ii) TS of the stele of a root of *Vicia faba* showing a complete 'ring' of vascular tissue, and production of secondary tissue inside the primary phloem (x 140)

primary phloem

vascular cambium

metaxylem

protoxylem

cells developing into secondary xylem

Cork formation follows the formation of vascular cambium. The cork cambium (phellogen) arises in the pericycle, just inside the central stele. It cuts off cells to the outside only by tangential divisions of the phellogen. These cells differentiate into cork tissue. The combined increase in thickness due to secondary vascular tissue and to cork tissue compresses the cortex and ruptures it. The whole of the cortex, with exodermis and epidermis, is sloughed off. Consequently a secondary thickened root will initially be substantially narrower than a similar root of the plant showing primary growth only.

Figure 3.10(i) Secondarily thickened root of *Vicia faba* (Broad Bean) (x 140)

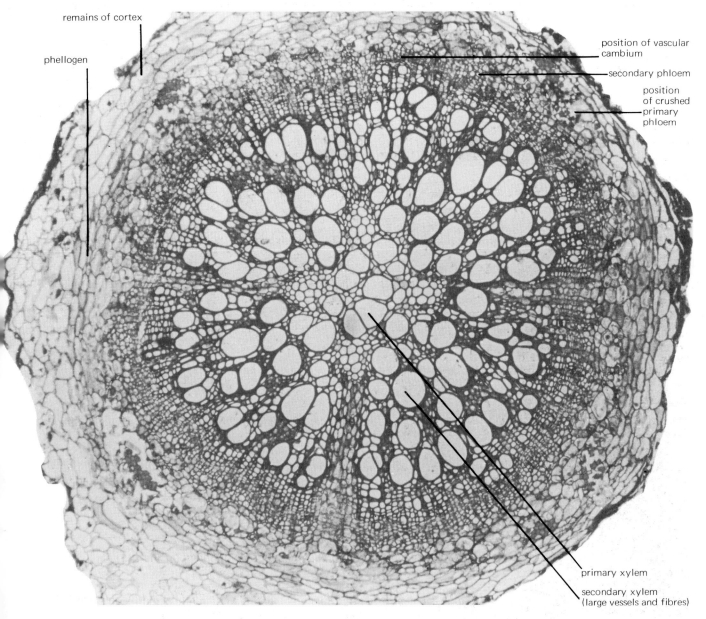

remains of cortex

phellogen

position of vascular cambium

secondary phloem

position of crushed primary phloem

primary xylem

secondary xylem (large vessels and fibres)

Section 4 : Cell division and reproduction

Introduction

Division of a cell is preceded by nuclear division. The nucleus contains *chromosomes* which determine and control all the activity of the cell by controlling protein and enzyme synthesis. In cell division separation of the 'daughter' nuclei is followed by the laying down of a wall between the two cells. The nucleus remains the same size throughout life, but appears large in relation to cell size in a newly formed cell.

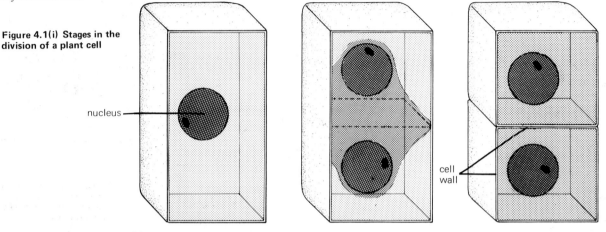

Figure 4.1(i) Stages in the division of a plant cell

nucleus

cell wall

Cell division may be stimulated in a number of ways, for example:
- *(a)* when the cell is so large that the nucleus is no longer adequate to regulate and control the activities of the cell;
- *(b)* by plant growth regulators, e.g. kinetin;
- *(c)* by postulated reproductive development regulators, e.g. florigen.

Sexual and asexual reproduction

Sexual reproduction involves the production and fusion of *gametes* (sex cells) to form a zygote.

The zygote then undergoes cell division, and the process is repeated continuously in the developing embryo. At each division the chromosomes of the daughter cells are identical to those of the cell from which they were derived. Thus in this type of division (called *mitotic division* or **mitosis**) the chromosomes are replicated (an exact copy is made) and one of each passes into each daughter nucleus.

In the production of gametes in flowering plants a reductive division (called **meiosis**) occurs in which the number of chromosomes i each gamete is half that in the ordinary cells. If such a reductive division did not occur in the life cycle the number of chromosomes would double at each fertilisation. (See the life-cycles on pages 63–64).

Any mechanism by which part of a plant is cut off and develops into an additional individual is asexual or vegetative reproduction. Such a process results in new individuals that are genetically identical to the parent. Only mitotic cell division is involved. In meiosis some variation occurs between the information of the chromosomes in each gamete. The process of fertilisation, whilst restoring the normal chromosome number in the *zygote*, adds further variation in the progeny.

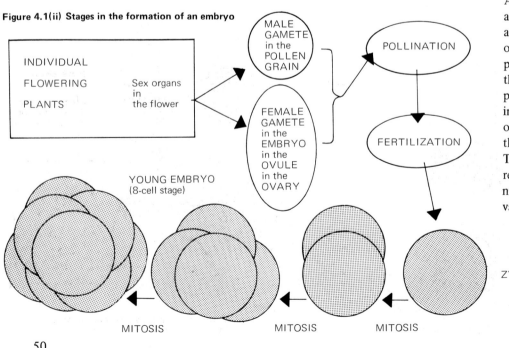

Figure 4.1(ii) Stages in the formation of an embryo

INDIVIDUAL FLOWERING PLANTS — Sex organs in the flower

MALE GAMETE in the POLLEN GRAIN

FEMALE GAMETE in the EMBRYO in the OVULE in the OVARY

POLLINATION

FERTILIZATION

YOUNG EMBRYO (8-cell stage)

MITOSIS   MITOSIS   MITOSIS

ZYGOTE

50

Mitosis

Mitosis is a process that takes about 60 minutes to complete. The diagrams and photographs here cover four crucial stages in a continuous process which divides chromosomes into identical groups, both of which are identical to the parent.

Figure 4.2(i–v) LS of root-tip cells of *Allium* sp. (Onion) showing the nucleus at various stages of mitosis, each accompanied by an explanatory line drawing (x 1250)

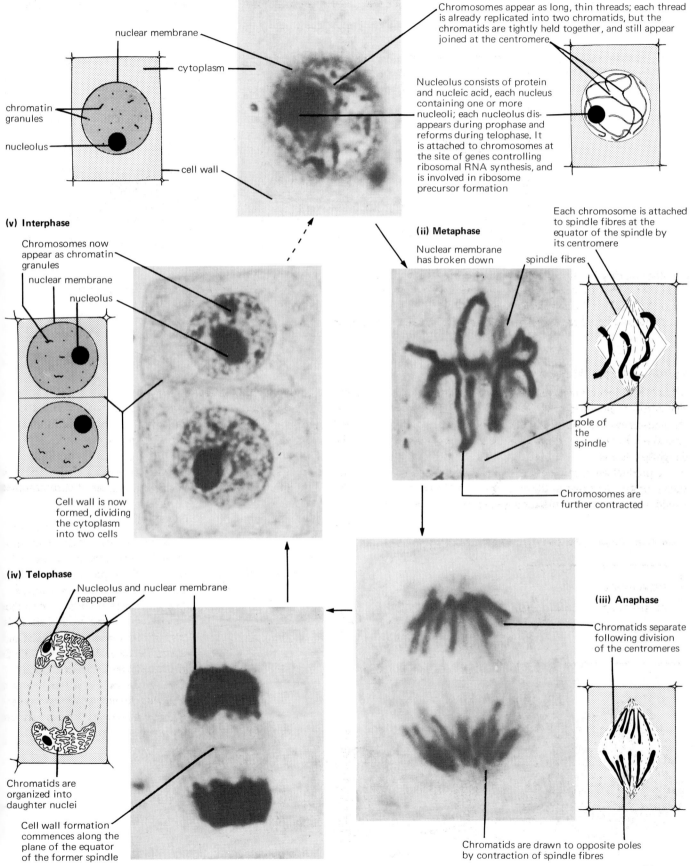

(i) Prophase

Chromosomes appear as long, thin threads; each thread is already replicated into two chromatids, but the chromatids are tightly held together, and still appear joined at the centromere.

Nucleolus consists of protein and nucleic acid, each nucleus containing one or more nucleoli; each nucleolus disappears during prophase and reforms during telophase. It is attached to chromosomes at the site of genes controlling ribosomal RNA synthesis, and is involved in ribosome precursor formation

nuclear membrane

cytoplasm

chromatin granules

nucleolus

cell wall

(v) Interphase

Chromosomes now appear as chromatin granules

nuclear membrane

nucleolus

Cell wall is now formed, dividing the cytoplasm into two cells

(ii) Metaphase

Nuclear membrane has broken down

Each chromosome is attached to spindle fibres at the equator of the spindle by its centromere

spindle fibres

pole of the spindle

Chromosomes are further contracted

(iv) Telophase

Nucleolus and nuclear membrane reappear

Chromatids are organized into daughter nuclei

Cell wall formation commences along the plane of the equator of the former spindle

(iii) Anaphase

Chromatids separate following division of the centromeres

Chromatids are drawn to opposite poles by contraction of spindle fibres

Chromosomes and chromosome numbers

Chromosomes are visible only at times of cell division, and then only after careful staining of the nucleus or by phase contrast microscopy. At *interphase* (the nucleus not involved in division) the contents of the nucleus show up as chromatin granules. The nucleus of each cell does contain a definite number of fine, coiled, thread-like chromosomes containing deoxyribonucleic acid (DNA). But at interphase they are mostly dispersed and committed to the production and despatch of messenger information, ribonucleic acid (RNA), to ribosomes in the cytoplasm. Cells of *Ranunculus acris* (Meadow Buttercup) contain 14 chromosomes; cells of *Taraxacum officinale* (Common Dandelion) contain 24 chromosomes; cells of *Ilex aquifolium* (Holly tree) contain 40 chromosomes; cells of *Homo sapiens* (Man!) contain 46 chromosomes.

This number of chromosomes is, in each case, the *'diploid'* or double number of chromosomes. Most individuals arise by sexual reproduction and subsequent mitotic cell division. The chromosomes in each cell are in pairs, one of each pair having come from each parent. In *meiosis*, which occurs in the sex organs within the flower during gamete formation, four cells are formed from each gamete mother-cell, each having half the chromosome number of the parent. Thus the process of meiosis establishes a single set of unpaired chromosomes (*haploid* number) in the pollen grains and in the egg cell of the embryo sac.

Meiosis also establishes *variation* between the gametes. The information within each gamete is different. This is achieved because:

(1)   there is independent assortment of chromosomes during meiosis; either one of a pair of chromosomes can enter the gamete with either chromosome of any other pair;

(2)   during meiosis individual chromatids break and rejoin during the first prophase; these are called chiasmata and result in exchange of DNA between individual chromosomes.

The crucial stages in the continuous process of meiosis are described below.

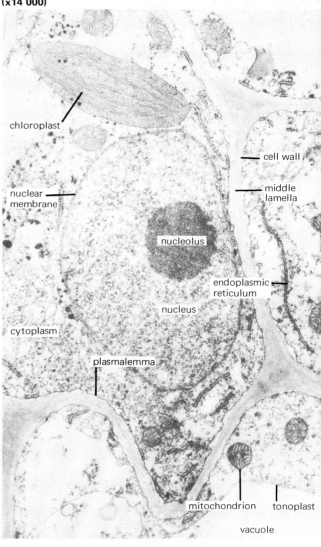

Figure 4.3(i) Electronmicrograph of part of a phloem parenchyma cell of *Nymphoides peltata* (Fringed Water Lily) showing nuclear material (x14 000)

Figure 4.3(ii) TS of a pollen mother cell in an anther of *Lilium* sp. (Lily) showing the nucleus at leptotene of Prophase I of meiosis (x 1000)

Meiosis

Meiosis can be seen as two, very much modified mitotic divisions. Prophase I is long and elaborate and is subdivided into stages (a)–(e). The chromosomes appear as single threads. Homologous chromosomes then pair up to form a *bivalent* that shortens and thickens. The chromosomes replicate themselves, but the two pairs of chromatids remain joined at their centromeres. Chiasmata occur. The mutual attraction between the chromosomes of the bivalent lapses, but they are held together by the chiasmata. Then a spindle forms (Metaphase I) and the bivalents move to the equator. In Anaphase I the members of each bivalent move to opposite poles, each pair of chromatids still held together by their centromeres.

Figure 4.3(iii) Drawings of 12 stages of meiosis

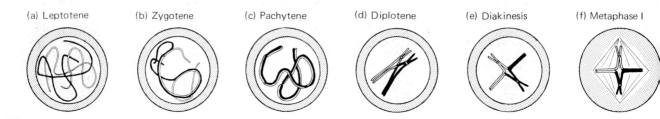

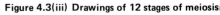

(a) Leptotene   (b) Zygotene   (c) Pachytene   (d) Diplotene   (e) Diakinesis   (f) Metaphase I

Figure 4.4(i) TS of pollen mother cells in an anther of *Lilium* sp. (Lily) showing nuclei at diakinesis of Prophase I and at Metaphase I (x 2000)

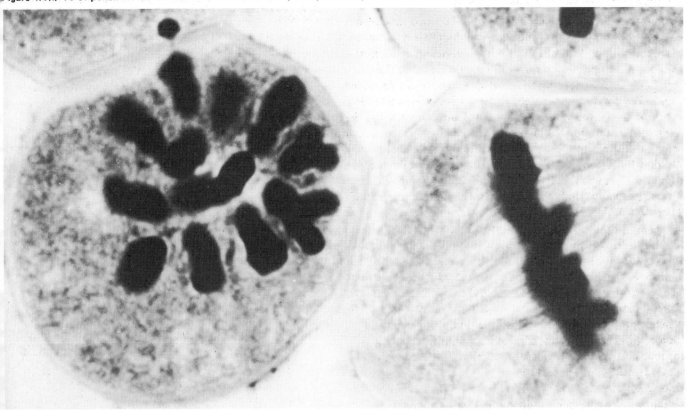

Figure 4.4.(ii) Nucleus now at Anaphase I (x 2000)

In the second meiotic division, stages (i)–(l) spindles are formed at right angles to the first one and the chromatids in each pair separate. Four nuclei are formed, each enclosing the haploid number of chromosomes. The cytoplasm is then divided too.

Figure 4.4(iii) Nucleus now at Telophase I (x 2000)

(g) Anaphase I

(h) Telophase I

(i) Interphase

(j) Metaphase II

(k) Anaphase II

(l) Telophase II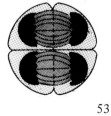

The flower and flowering

The flower is the organ of sexual reproduction, and is usually an hermaphrodite structure, but it may be unisexual with separate ma[le] and female flowers on the same plant (monoecious, e.g. *Corylus avellana* (Hazel), or with separate male and female plants (dioeciou[s] e.g. *Mercurialis perennis* (Dog's Mercury)).

The flower develops at the tip of a shoot, and the parts of the flower may be seen as modified leaves. But we are ignorant of the evo[?]lutionary history of the flower (flowers are ephemeral structures, rarely preserved in the fossil record), and it is a matter of conject[?]ure whether comparison can be made between the flower of modern angiosperms and the spore-producing structures of other trache[?] phytes (meaning plants with vascular tissue, e.g. cones of *Selaginella* spp. (Club Moss), cones of *Pinus sylvestris* (Scots Pine); see p.6[?] However, the point of transition from vegetative growth to flowering in plant development has been investigated. The external temp[?] erature and the change in day length have proved to be environmental factors of key importance. After a certain stage in vegetative growth (i.e. when a certain number of nodes has been produced) a condition of *'ripeness to flower'* is achieved. *Day-neutral* varietie[s] will then flower, e.g. *Lycopersicum esculentum* (Tomato). Such plants often continue vegetative and reproductive growth simultan-eously. Other plants then require exposure to days of a particular length before they will flower *(photoperiodism)*. For example, many plants of temperate regions will flower only in the long days of summer and are called *long-day plants*, e.g. *Avena sativa* (Oats[?] Other plants require long dark periods and this group includes many plants indigenous to regions of low latitude north and south of the equator where day length never exceeds more than 14 hours at any season of the year. An example of a *short-day plant* is *Glyci*[?] *soja* (Soya Bean).

Flowers may occur solitarily on a stem, or in groups (an *inflorescence)* on the same main stem. The parts of each flower are borne i[n] whorls on the expanded tip *(receptacle)* of the flower stalk *(pedicel)*.

The *sepals* (collectively called the calyx) are usually green and small. They enclose the flower in the bud.

The *petals* (collectively called the *corolla)* are often coloured and conspicuous, and may attract insects to the flower.

The *stamens (androecium* – derived from the Greek for 'man' and 'house') are the male parts of the flower, consisting of *anthers* (containing pollen grains) and a *filament* (stalk).

The *carpels* (*gynoecium* – derived from the Greek for 'woman' and 'house') may be one or many, separate or joined. Each consists [of] an *ovary* (containing ovule/s), *stigma* (receptive surface for pollen), and a *style* (connecting ovary and stigma).

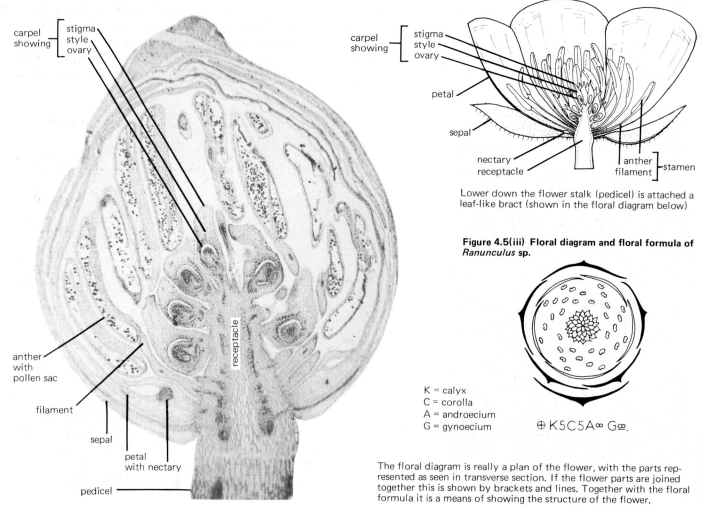

Figure 4.5(i) LS of a flower bud of *Ranunculus* sp. (Buttercup) (x 25)

carpel showing — stigma / style / ovary

anther with pollen sac

filament

sepal

petal with nectary

pedicel

receptacle

Figure 4.5(ii) Half-flower of *Ranunculus repens* (Buttercup)

carpel showing — stigma / style / ovary

petal

sepal

nectary receptacle

anther / filament — stamen

Lower down the flower stalk (pedicel) is attached a leaf-like bract (shown in the floral diagram below)

Figure 4.5(iii) Floral diagram and floral formula of *Ranunculus* sp.

K = calyx
C = corolla
A = androecium
G = gynoecium

⊕ K5C5A∞ G∞.

The floral diagram is really a plan of the flower, with the parts rep-resented as seen in transverse section. If the flower parts are joined together this is shown by brackets and lines. Together with the floral formula it is a means of showing the structure of the flower.

Figure 4.6(i) Half-flower and floral formula of *Lilium* sp. (Lily)

Figure 4.6(ii) TS of flower bud of *Lilium* sp. (x 40)

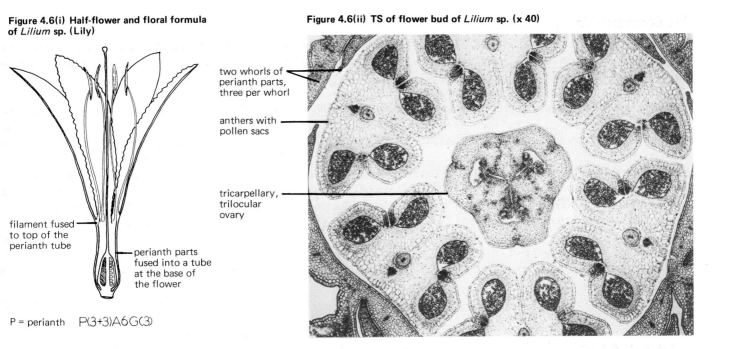

filament fused to top of the perianth tube

perianth parts fused into a tube at the base of the flower

P = perianth   P(3+3)A6G(3)

two whorls of perianth parts, three per whorl

anthers with pollen sacs

tricarpellary, trilocular ovary

There is a similar basic structure to most flowers, but there is enormous variation too. Many dicotyledons have flower parts in fives but the Cruciferae (e.g. *Sinapis arvensis* (Charlock)) have their parts in fours, and the Poppy family in twos. *Ranunculus* sp. (Buttercup) has a very large number of stamens and carpels, but many plants have small and regular numbers of both stamens and carpels. Flower parts are not always symmetrically arranged about any axis as in regular or *actinomorphic* flowers. The Leguminosae (e.g. *Lupinus polyphyllus* (Garden Lupin)) and the Labiatae (e.g. *Lamium album* (White Deadnettle)) are examples with *zygomorphic* flowers, having only one plane of symmetry. Also the flower parts may be joined into tubes, as is both calyx and corolla of White Deadnettle. If the sepals and petals are indistinguishable they are called the *perianth*. The perianth may be petaloid as in *Lilium* spp. (Lilies). This is the case in many families of monocotyledons, where flower parts occur in threes. Many flowers have a conical receptacle like that of *Ranunculus* spp.; in others the receptacle is saucer- or flask-shaped, and for some the ovary is *'inferior'* being completely embedded in receptacle tissue. The various ways in which flowers are grouped together give rise to different forms of inflorescence, all classified according to whether the youngest flower occurs at the apex or at the base. Probably the least wasteful and therefore most efficient inflorescence is as in the head or *capitulum* shown by Compositae, where flowers are clustered tightly together.

Figure 4.6(iii) LS of capitulum of *Taraxacum* sp. (Dandelion) (x 5)

Figure 4.6(iv) TS of capitulum of *Scabiosa* sp. (Scabious) (x 50)

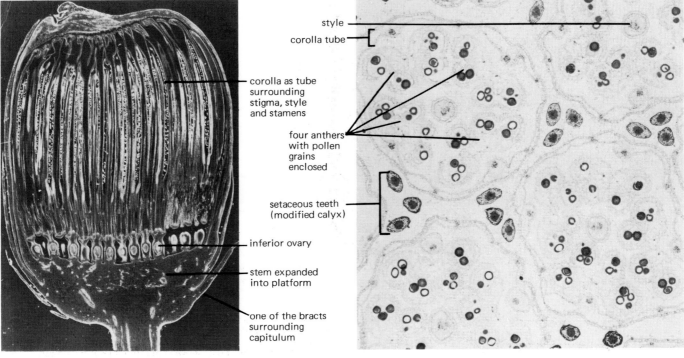

corolla as tube surrounding stigma, style and stamens

inferior ovary

stem expanded into platform

one of the bracts surrounding capitulum

style

corolla tube

four anthers with pollen grains enclosed

setaceous teeth (modified calyx)

The gynoecium

The gynoecium (female part of the flower) consists of a *carpel* or carpels which may be free-standing or fused together in various ways. Most carpels sit upon the receptacle tissue (ovary *superior*), but in some cases the ovary is embedded in the receptacle (ovary *inferior*). The evolutionary origin *(phylogeny)* of the carpel may be from a fertile leaf, the margins of which bore ovules.

Figure 4.7(i) Diagrammatic representation of probable evolutionary development of present-day *Lilium* (Lily) carpels from a fertile leaf

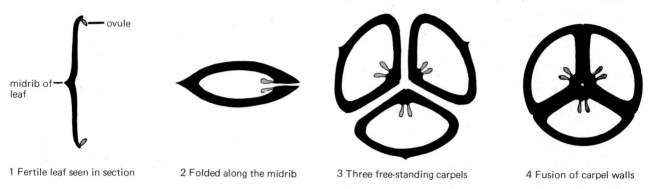

1 Fertile leaf seen in section   2 Folded along the midrib   3 Three free-standing carpels   4 Fusion of carpel walls

The ovule consists of the *nucellus* (parenchyma cells) which is surrounded by *integuments* (when these develop) and attached to the *placenta* of the ovary by a stalk (the *funiculus*). Within the nucellus is the *megaspore-mother-cell*.

Figure 4.7(ii) Stages in the ontogeny of the ovule

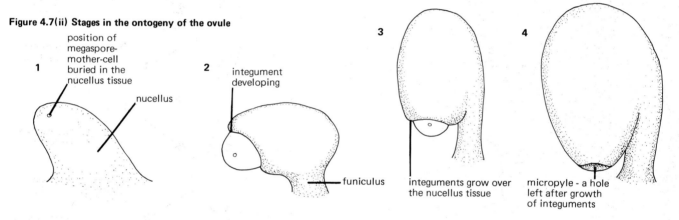

Figure 4.7(iii) TS of part of the ovary of *Lilium* sp. showing one ovule in LS (x 270)

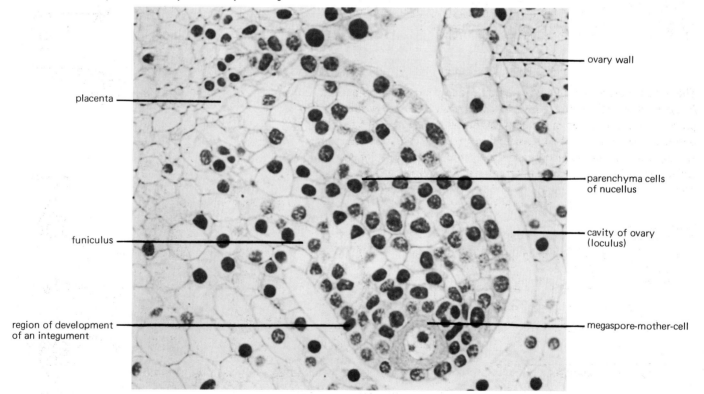

The female gamete is the *egg cell*, which develops within the *embryo sac*. The embryo sac develops from the megaspore-mother-cell. Numerous recent studies have discovered surprising diversity in the details of this process within different angiosperm species. In *Fritillaria* spp. and *Lilium* spp. megasporogenesis and megagametogenesis are as shown in the diagram below:

Figure 4.8(i) LS of ovule of *Fritillaria* sp. (Fritillary) showing the mature embryo sac (× 130)

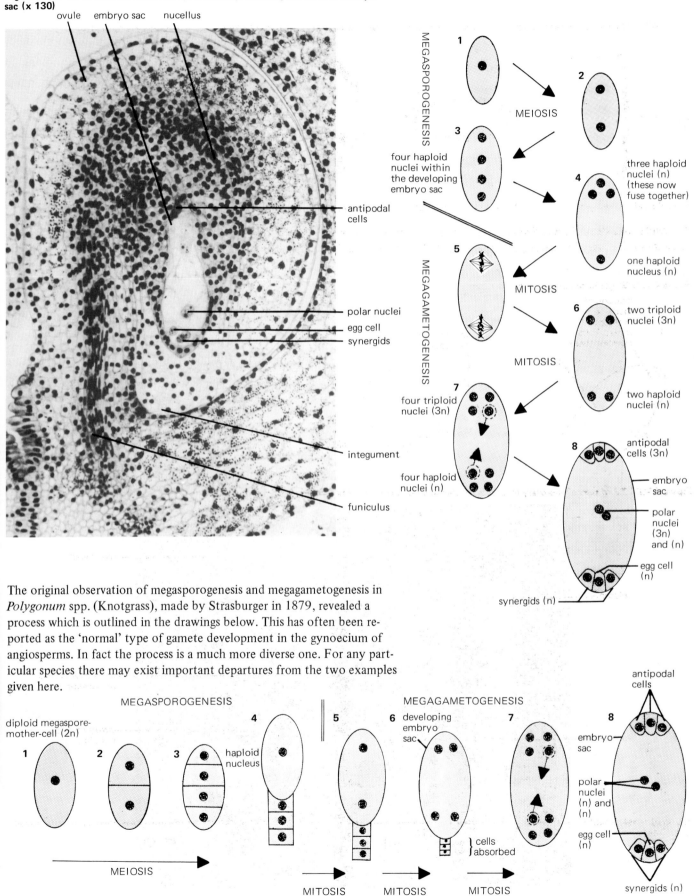

The original observation of megasporogenesis and megagametogenesis in *Polygonum* spp. (Knotgrass), made by Strasburger in 1879, revealed a process which is outlined in the drawings below. This has often been reported as the 'normal' type of gamete development in the gynoecium of angiosperms. In fact the process is a much more diverse one. For any particular species there may exist important departures from the two examples given here.

57

The androecium and the male gamete

The androecium (male part of the flower) consists of *stamens*, which are upgrowths from the receptacle. Each stamen consists of an *anther*, containing four pollen sacs (microsporangia), and a *filament*.

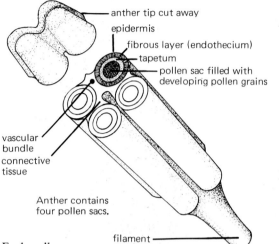

anther tip cut away
epidermis
fibrous layer (endothecium)
tapetum
pollen sac filled with developing pollen grains

vascular bundle
connective tissue

Anther contains four pollen sacs.

filament

Each pollen sac has a wall several cell layers thick; the innermost layer *(tapetum)* provides food for the developing *pollen grains (microspores)*. The *microspore-mother-cells* undergo meiosis, so that each cell produces four pollen grains in a tetrad. Pollen grain formation goes on in a nutritive liquid secreted by the tapetum. Pollen grains may be arranged in various ways in the tetrads, e.g:

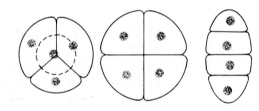

Figure 4.9(i) TS of part of the anther of *Lilium* sp. (Lily) showing high-power detail of part of a pollen sac (microsporangium) (x 250)

microspore-mother-cell   tapetum   cells of wall layers   epidermis of anther

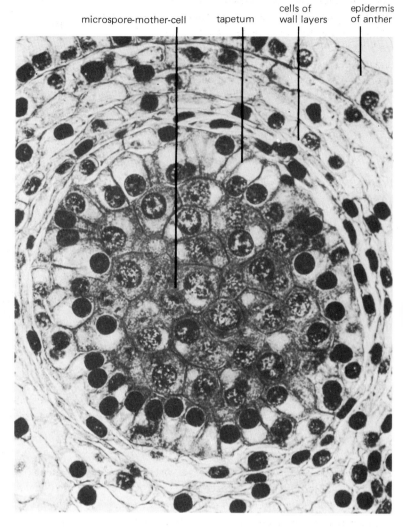

Figure 4.9(ii) TS through part of *Lilium* sp. pollen sac, showing nuclei at metaphase of meiosis (x 200)

Figure 4.9(iii) TS through part of *Lilium* sp. pollen sac, showing a tetrad at late telophase (x 750)

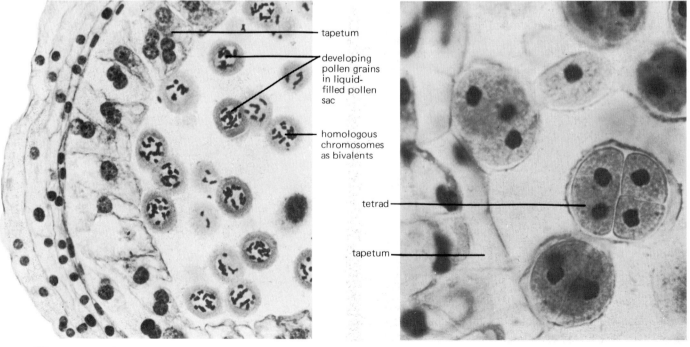

tapetum

developing pollen grains in liquid-filled pollen sac

homologous chromosomes as bivalents

tetrad

tapetum

As the anther matures a fibrous layer of cells (the endothecium) appears below the epidermis, with U-shaped thickenings involving all but the outer facing wall of each cell. As these cells dry out, the tension created ruptures the thin-walled cells (the stomium) at the junction of the pollen sacs. A longitudinal split exposes the pollen grains. (The majority of anthers dehisce in this way.)

Figure 4.10(i) TS through the ripe anther of *Lilium* sp. (Lily) near the base, at a point where the filament is not fused with the connective (x 110)

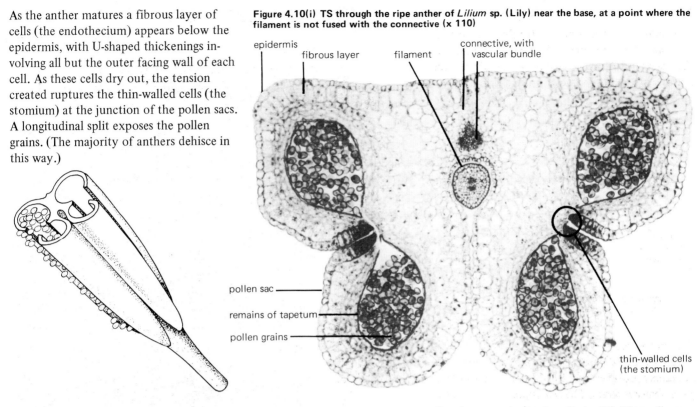

epidermis
fibrous layer
filament
connective, with vascular bundle
pollen sac
remains of tapetum
pollen grains
thin-walled cells (the stomium)

The mature pollen grain has a haploid nucleus which divides to form a *generative cell* and a *vegetative nucleus* (see p.60). The pollen grain is surrounded by a thin cellulose wall, the *intine*, and an outer, usually sculptured wall, the *exine*. The pollen grain wall has thin areas (apertures) where growth of the pollen tube can occur. Pollen from monocotyledons is often oval shaped with one aperture; that from dicotyledons usually has three apertures.

Analysis of the different sculpturings on the surface of pollen grains is of great importance to:

(1) understanding the history of floras all over the world — by means of quantitative pollen analysis of recent geological layers it is possible to determine the species of plants that existed when no other relic of them has been preserved;
(2) the study of allergic diseases — pollen grains are one of the most important causes of allergic disease of the respiratory tract;
(3) research on honey plants — the pollen collected by bees can be identified. Pollen shows a chemical composition of the following order: proteins, 7–26%; carbohydrates (starch, later converted to sugar), 24–48%; fats, 1–14%; ash (i.e. minerals), 1–5%; and water 7–16%.

Figure 4.10(ii) Pollen grain of *Gossypium* sp. (Cotton), seen by scanning electron microscope (x 1400)

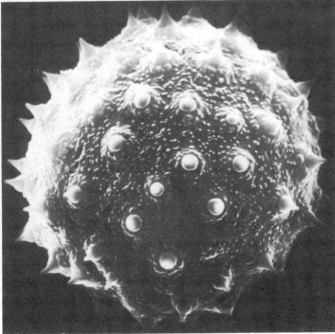

Figure 4.10 (iii) Pollen grain of *Lilium* sp. (x 1500)

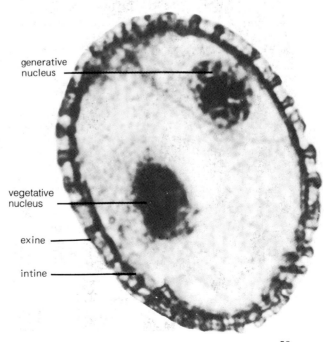

generative nucleus
vegetative nucleus
exine
intine

Pollination

Pollination is the transfer of pollen from the stamens to the stigma. When the stamens and stigma are in the same flower the process is described as *self-pollination*. When they are in different flowers, particularly on separate plants, this is described as *cross-pollination*. The important consequence of cross-pollination is a greatly enhanced opportunity for variation in genetic material of the resulting progeny. In true cross-pollination the chromosomes are a new mixture from the chromosomes of the two parents. In self-pollination the progeny have chromosome recombinations of the one parent's chromosomes.

Figure 4.11(i) LS of the stigma and part of the style of *Oenothera* sp. (Evening Primrose), showing pollen-tube growth (x 180)

As the pollen grain has no power of independent movement, pollination is usually by means of insects *(entomophily)* or by means of wind *(anemophily)*. Plants with flowers so pollinated have distinctive features, as listed in the table below:

Entomophilous flowers	Anemophilous flowers
1 Insects are attracted by scent or by colour, and often by both. Nectaries may be present, formed from a part of the flower adapted to secrete nectar (dilute sugar solution).	1 Produce an enormous amount of pollen, most of which is lost since the stigma is a small target and the wind is random in its operations.
2 Pollen is often sticky and oily, sufficiently so to facilitate temporary attachment to the body surface of the insect.	2 Pollen is light, small, and smooth.
3 Stamens may be protected from heavy rain and from insects other than those able to pollinate by the shape, position, and sometimes the fusion of the parts of the calyx and corolla.	3 Flowers are often unisexual, with an excess of male flowers.
4 Insect visitors for which flowers may show specific adaptations include beetles, flies, butterflies and moths, sawflies, wasps, ants, and bees.	4 Stamens and stigmas project clear of perianth or bracts. The anthers are versatile (filament attached near the middle, allowing movement) and the stigmas are often feathery. Anemophily found mostly in plants that are gregarious.

Once the pollen grain has reached the stigma the following changes may occur:

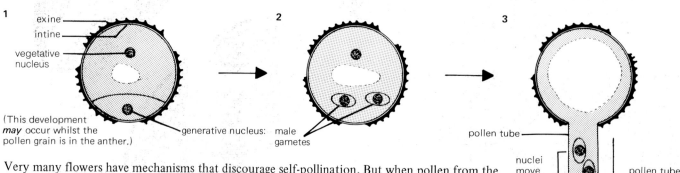

Very many flowers have mechanisms that discourage self-pollination. But when pollen from the same individual does germinate on the style *physiological incompatibility* may prevent pollen tube growth. A simple example is when the pollen grain has a totally different osmotic potential from that of the stigma cells: the pollen tube ruptures or collapses, and in either case fails to achieve fertilization.

60

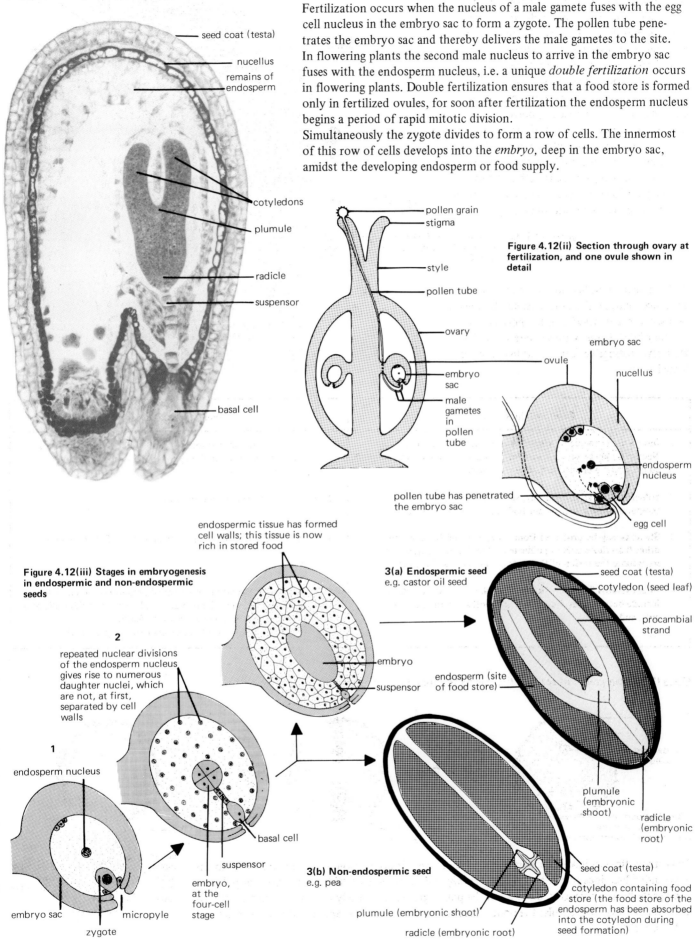

Figure 4.12(i) LS to show detail of embryo in embryo sac of *Capsella bursa-pastoris* (Shepherd's Purse) (x 300)

seed coat (testa)

nucellus

remains of endosperm

cotyledons

plumule

radicle

suspensor

basal cell

Fertilization

Fertilization occurs when the nucleus of a male gamete fuses with the egg cell nucleus in the embryo sac to form a zygote. The pollen tube penetrates the embryo sac and thereby delivers the male gametes to the site. In flowering plants the second male nucleus to arrive in the embryo sac fuses with the endosperm nucleus, i.e. a unique *double fertilization* occurs in flowering plants. Double fertilization ensures that a food store is formed only in fertilized ovules, for soon after fertilization the endosperm nucleus begins a period of rapid mitotic division.

Simultaneously the zygote divides to form a row of cells. The innermost of this row of cells develops into the *embryo*, deep in the embryo sac, amidst the developing endosperm or food supply.

pollen grain
stigma
style
pollen tube
ovary
embryo sac
male gametes in pollen tube

Figure 4.12(ii) Section through ovary at fertilization, and one ovule shown in detail

ovule

embryo sac

nucellus

endosperm nucleus

pollen tube has penetrated the embryo sac

egg cell

Figure 4.12(iii) Stages in embryogenesis in endospermic and non-endospermic seeds

endospermic tissue has formed cell walls; this tissue is now rich in stored food

3(a) Endospermic seed e.g. castor oil seed

seed coat (testa)
cotyledon (seed leaf)
procambial strand

embryo

suspensor

endosperm (site of food store)

2

repeated nuclear divisions of the endosperm nucleus gives rise to numerous daughter nuclei, which are not, at first, separated by cell walls

1

endosperm nucleus

basal cell

suspensor

embryo, at the four-cell stage

embryo sac

micropyle

zygote

plumule (embryonic shoot)

radicle (embryonic root)

3(b) Non-endospermic seed e.g. pea

plumule (embryonic shoot)

radicle (embryonic root)

seed coat (testa)

cotyledon containing food store (the food store of the endosperm has been absorbed into the cotyledon during seed formation)

61

The seed

The seed (e.g. the pea) develops from the fertilized ovule and contains an *embryonic plant* and a *food store.* The seed has one scar, being the point of attachment to the placenta. The fruit (e.g. the pea pod) develops from the ovary and contains the seed or seeds. The fruit has two scars, the scars left by the withered style and by the attachment to the receptacle. A *true fruit* (e.g. tomato) is made of ovary alone; a *false fruit* (e.g. apple) contains receptacle tissue with the ovary wall. There is biological advantage in having a mechanism by which seeds are dispersed away from the parent plants, yet many seeds have no such mechanism and are dispersed accidentally by the careless actions of animals (including man) that eat many seeds. In other fruits the ovary wall breaks open violently, scattering seeds, and many seeds have structures which aid dispersal by wind or by animals.

The seed is a form which survives the unfavourable season *(perennation).* It is a form in which dispersal of the plant can occur. Many seeds have a *dormant* period and may only germinate after specific external conditions have been met (e.g. cold treatment). When conditions are ideal for germination not all seeds do so immediately; there is some staggering of the crop in the wild state. This feature has been bred out of cultivated plants over many years.

Figure 4.13(i) LS through fruit of *Zea mays* (Maize) (x 15)

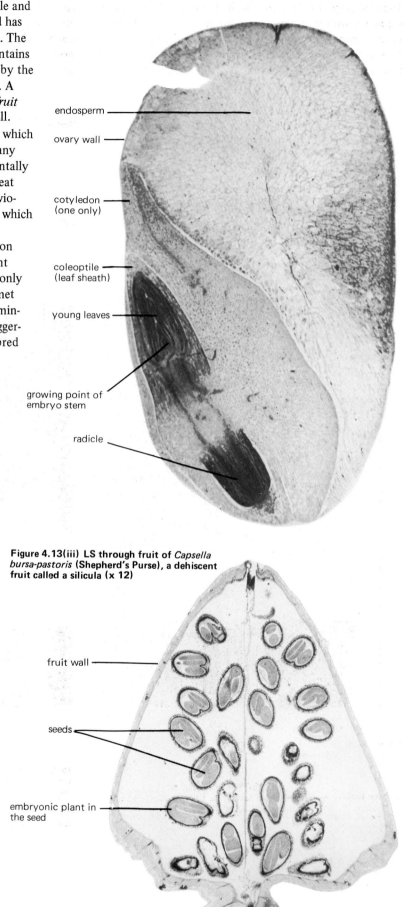

endosperm

ovary wall

cotyledon (one only)

coleoptile (leaf sheath)

young leaves

growing point of embryo stem

radicle

Figure 4.13(ii) LS through seed of *Helianthus* sp. (Sunflower) (x 12)

cotyledon with food store

seed coat (testa)

plumule

radicle

Figure 4.13(iii) LS through fruit of *Capsella bursa-pastoris* (Shepherd's Purse), a dehiscent fruit called a silicula (x 12)

fruit wall

seeds

embryonic plant in the seed

The seed in the life-cycle

The seed plays a unique role in the life-cycle of the flowering plant. This can be appreciated by comparing the life-cycles of flowering plants with those of the present-day survivors of the flowering plants' ancestors.

In the flowering plants, reproducing by seeds, each new individual begins life as a multicellular embryo equipped with a protective coat and a food supply. The chances of survival for seeds are therefore higher than those for the spores produced by the non-flowering plants, which are less protected, and are more dependent on an external food supply.

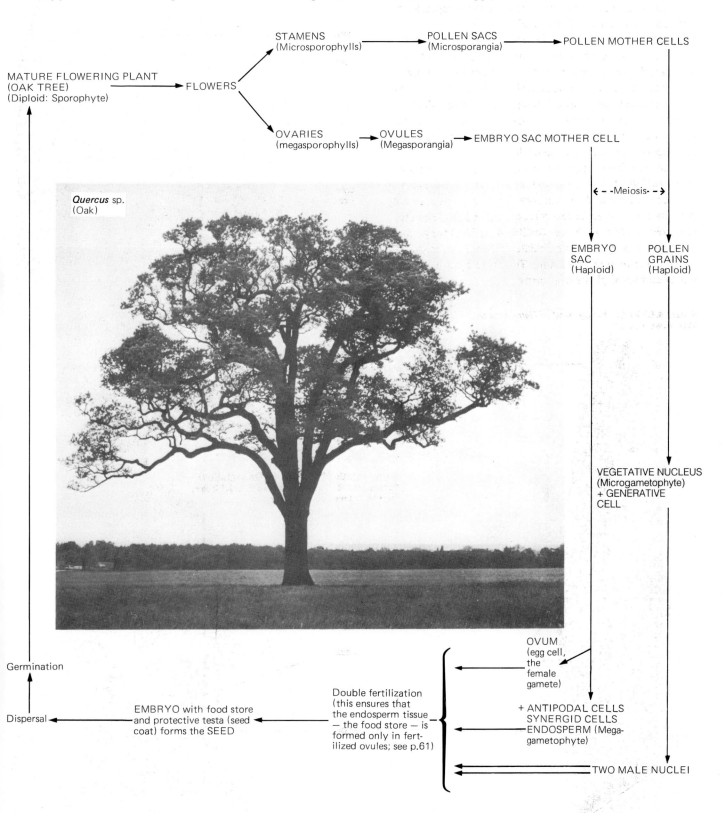

MATURE FLOWERING PLANT
(OAK TREE)
(Diploid: Sporophyte) → FLOWERS

STAMENS
(Microsporophylls) → POLLEN SACS
(Microsporangia) → POLLEN MOTHER CELLS

OVARIES
(megasporophylls) → OVULES
(Megasporangia) → EMBRYO SAC MOTHER CELL

Quercus sp.
(Oak)

← – -Meiosis- – →

EMBRYO
SAC
(Haploid)

POLLEN
GRAINS
(Haploid)

VEGETATIVE NUCLEUS
(Microgametophyte)
+ GENERATIVE
CELL

OVUM
(egg cell,
the
female
gamete)

+ ANTIPODAL CELLS
SYNERGID CELLS
ENDOSPERM (Mega-
gametophyte)

TWO MALE NUCLEI

Double fertilization
(this ensures that
the endosperm tissue
— the food store — is
formed only in fert-
ilized ovules; see p.61)

EMBRYO with food store
and protective testa (seed
coat) forms the SEED

Germination

Dispersal

Haploid and diploid

Comparison of life-cycles of flowering plants and more primitive plants also establishes the gradual reduction of the haploid phase. Organisms in the diploid condition have the opportunity of showing greater genetic variation, and therefore more varied control of themselves.

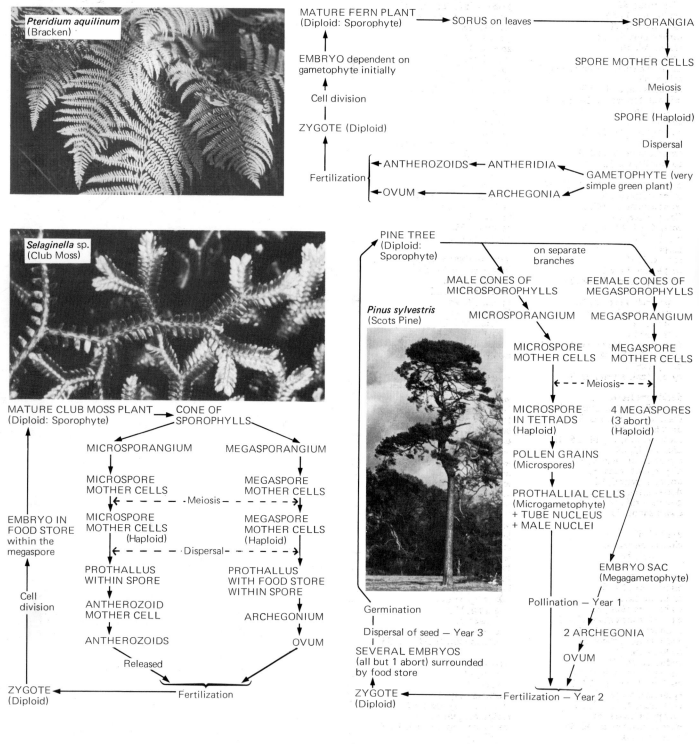

Pteridium aquilinum (Bracken)

MATURE FERN PLANT (Diploid: Sporophyte) → SORUS on leaves → SPORANGIA

EMBRYO dependent on gametophyte initially

Cell division

ZYGOTE (Diploid)

SPORE MOTHER CELLS

Meiosis

SPORE (Haploid)

Dispersal

Fertilization { ANTHEROZOIDS ← ANTHERIDIA ← GAMETOPHYTE (very simple green plant)

OVUM ← ARCHEGONIA

Selaginella sp. (Club Moss)

MATURE CLUB MOSS PLANT (Diploid: Sporophyte) → CONE OF SPOROPHYLLS

MICROSPORANGIUM   MEGASPORANGIUM

MICROSPORE MOTHER CELLS   MEGASPORE MOTHER CELLS
— — — Meiosis — — —

MICROSPORE MOTHER CELLS (Haploid)   MEGASPORE MOTHER CELLS (Haploid)
— — Dispersal — —

PROTHALLUS WITHIN SPORE   PROTHALLUS WITH FOOD STORE WITHIN SPORE

ANTHEROZOID MOTHER CELL   ARCHEGONIUM

ANTHEROZOIDS   OVUM

EMBRYO IN FOOD STORE within the megaspore

Cell division

ZYGOTE (Diploid) ← Released   Fertilization

PINE TREE (Diploid: Sporophyte)

on separate branches

Pinus sylvestris (Scots Pine)

MALE CONES OF MICROSPOROPHYLLS   FEMALE CONES OF MEGASPOROPHYLLS

MICROSPORANGIUM   MEGASPORANGIUM

MICROSPORE MOTHER CELLS   MEGASPORE MOTHER CELLS
— — Meiosis — — →

MICROSPORE IN TETRADS (Haploid)   4 MEGASPORES (3 abort) (Haploid)

POLLEN GRAINS (Microspores)

PROTHALLIAL CELLS (Microgametophyte) + TUBE NUCLEUS + MALE NUCLEI

EMBRYO SAC (Megagametophyte)

Germination

Dispersal of seed — Year 3

SEVERAL EMBRYOS (all but 1 abort) surrounded by food store

Pollination — Year 1

2 ARCHEGONIA

OVUM

ZYGOTE (Diploid) ← Fertilization — Year 2

Index